THE GAS PIPE NETWORKS

A HISTORY OF COLLEGE RADIO 1936-1946

The Brown Network control room in 1940. Left to right: Peter Thorpe, John Bailey, George Abraham and David Borst.

THE
GAS PIPE NETWORKS
A History of College Radio 1936-1946

by

Louis M. Bloch Jr.

Bloch and Company
Cleveland, Ohio

Published by Bloch and Company
P.O. Box 18058
Cleveland, Ohio 44118

Printed in the United States of America
by Lake City Litho Service Co., Wickliffe, Ohio

Typesetting by Meredith Graphics, Inc., Euclid, Ohio

This book is dedicated to George Abraham and David Borst, founders of "The Gas Pipe Networks" and to one and a half million students who have participated in college radio since 1936.

PREFACE

In the United States over 1,000 colleges have campus radio stations operated by students. Today these college stations operate as wired-wireless, closed circuit cable and FM. Some are educational, some commercial. College radio is the parallel of college journalism providing training in programming, announcing, engineering and business.

In the past 45 years over 1,500,000 individuals have participated in college radio. Many radio and television executives gained their early experience on campus stations. College radio has spread throughout the world. Almost every college and university in Australia and New Zealand has a campus station. Among other areas where stations operate are Canada, Great Britain and Latin America.

I had the great privilege of participating in the development of college radio from its very beginning at Brown University in 1936. In those days we had no idea that our inter-communication and miniature broadcasting system would develop into the greatest training ground for the radio industry and that it would enrich the lives of so many people.

This book tells the story of my involvement in college radio, affectionately called "The Gas Pipe Networks". Also related in this book is the story of college radio's founders, George Abraham and David Borst, both engineers. The college station was George's dream while much of the design and construction was by Dave with George participating. To this year, 1980, George and Dave are still active in The Intercollegiate Broadcasting System, the international, non-profit association of college radio stations.

LMB

CONTENTS

CHAPTER I

MY INTRODUCTION TO
"THE GAS PIPE NETWORK"

My arrival at Brown University as a freshman in September, 1936 changed the course of my life almost immediately as I was, after only a month, introduced to "The Gas Pipe Network". For the next ten years my major interest was to be college radio although I did not realize that at the time.

My dormitory, Hegeman Hall was a newer building, about twenty years old in 1936. Some older Brown buildings dated from the Revolution and University Hall housed Lafayette's troops during the Revolution. Brown is truly an Ivy League college with a great tradition and many distinguished alumni. Henry Merritt Wriston, a famous scholar, was then about to become President of the University. It was in this citadel of learning that I started my freshman year with my roommate, Alexander Franklin Black of New Rochelle, N.Y.

I had obtained my amateur radio license the year before, during my senior year at Cleveland Heights High School. Although radio was one of my interests and although I had constructed my receiver and transmitter, I never was an expert in electronics. I learned that George Abraham and David Borst, both freshmen, had linked their rooms in Caswell and Littlefield Halls, two dormitories about a quarter of a mile apart, by wire and had set up an intercommunication system between them. It was at this point that I contacted.them and was invited to participate in the Net. Dave Borst ran a line to my room and connected it to my radio receiver. Also, he connected a microphone which I could use for inter-communication. That small Net continued and very soon in the first semester of our freshman year we had many outlets all over the campus. Not only was I able to communicate with other participating freshmen but soon George and Dave added a second line so that the net could also operate as a broadcasting system. On special occasions programs were broadcast from the master control in George Abraham's room. We had special music programs and occasional interviews. The most important "first" of college radio occurred at that time. Henry Merritt Wriston was to be inaugurated as President of the University but the hall which was to be used on campus was too small to accommodate the students. George obtained permission to broadcast that event and that broadcast turned out to be the first major broadcast of college radio. Our audience was large and interested.

At first the Brown broadcasting and communication systems consisted of wire lines connected to the radios of participating students on campus and this complicated arrangement required a crew to string the lines, make the connections and maintain the system. As more and more students became interested, lines were strung to all parts of the campus. Dave was in charge of everything involved in engineering and in the installation and maintenance of the lines and soon had a crew of forty students helping him. Truly this was a network of lines tying the campus together and the press referred to our station as "The Brown Network". It was difficult for the college administration to comprehend what was going on and they had only a vague idea of what George and Dave were up to. And so it was, back in my freshman year, that college radio got its start.

In those days it was the custom for the sophomores to form a vigilance committee, a self appointed and unauthorized committee to keep the freshmen in their place. In past years this had been done by hazing, but in the year 1936 at Brown little did the sophomores know that the freshmen had their own communication system between most of the dormitories. They did know of our broadcasting to the campus but Line One, a line for communications was reserved only for freshmen, and for some unknown reason the sophomores involved in the vigilance committee never knew about Line One. As the vigilance committee moved from room to room and from building to building they were bombarded with water bombs, balloons filled with water thrown from the rooms of the freshmen, down the stairwells and also from the roofs of the buildings. Their every movement about campus was reported and after that disaster for the vigilance committee, they were never heard from again.

Our new president, Henry Wriston, while walking on campus inquired as to what all of those students were doing on the roofs of the buildings and when he was informed that they were just stringing and repairing lines for "The Brown Network" he dropped the subject and continued his walk without any further questions in that regard. The administration, it seems, had decided to tolerate us in the belief that this was just another type of student prank which would soon be forgotten. President Wriston did not see the seven lines running over the buildings tying the campus together originating from George's room. He did not interfere with "The Brown Network", nor did the college administration.

When I became a member of Pi Lambda Phi fraternity in my sophomore year I moved to another dormitory, Caswell Hall, with my new roommate Jerome Frank Strauss, Jr., now a prominent doctor in Chicago. My radio was again attached to the network lines. More and more students became interested in being tied into "The Brown Network" and George, in addition to broadcasting from his room, set up remote transmission points around the campus, including the Music Building. This opened up the vast classical music collection of the music department to us.

Through George's salesmanship "The Brown Network" became recognized as an extra-curricular activity and we were granted a room in the student union building, Faunce House. George, however, continued to maintain a broadcasting outlet in his room in Slater Hall. A formal organization was established and Dave and his engineering crew built a broadcasting studio with complete facilities including sound proof studios and a central control room with thirty outlets in dormitories and fraternities where programs could originate. A new transmission system was developed using low power limited to two watts. Programs were carried over 30,000 feet of wire strung through steam tunnels and over the roofs of buildings into the dormitories and fraternity houses. The power was so small that broadcasts could only be heard within the buildings covered, but special receivers were not required and programs could be tuned in at 570 on the radio dial. Since the Network did not radiate beyond the buildings, the station needed no license. In some cases the transmitter was coupled into the heating system of the buildings, or the electric lighting system, or in other cases into small remote transmitters located in some of the buildings.

The first "gas pipe" station in George Abraham's room, circa 1937. Left to right: David Borst, Joseph Parnicky, George Abraham, unidentified.

George Abraham broadcasting over the first "gas pipe" station. Photo by Frank Colby in the Christian Science Monitor ©1939 TCSPS.

CHAPTER II

THE BROWN NETWORK
COMES OF AGE

With the approval of The Brown Network as an established extra-curricular activity, many campus organizations cooperated with us, including the Glee Club, the dramatic group and the Brown Daily Herald, the campus newspaper, which featured our activities and programs on a regular basis.

In 1939 I became Business Manager succeeding Peter Thorpe. With Pete's help we developed a brochure entitled

"The Brown Network, America's First Collegiate Broadcasting System In One of New England's Key Cities." Rates were listed for Class A and Class B radio time, Class A being prime time, 8 p.m. to midnight and all day Sunday, and Class B including all other hours. Class A time rates were $50.00 per hour, $30.00 per half hour and $18.00 per quarter hour with $3.50 charged for one minute spots. Quantity discounts were allowed. Class B time rates were $30.00 per hour, $18.00 per half hour and $11.00 per quarter hour. One minute spots for Class B time were $2.50, all with quantity discounts.

Up to this point each of us had financed the station out of his own pocket. This was especially true of Dave and George who had financed the original station. I was able to obtain a small amount of advertising from local merchants but money was tight and collections difficult as we were at the end of the great depression. I had a major collection problem with a local men's store, but George solved the problem. He purchased a jacket at the store and notified the store owner that he had paid The Brown Network to cancel the bad debt of the store. George was a great diplomat and succeeded in solving some of our most difficult problems in a most unusual way.

On November 3, 1939, a distinguished visitor addressed our chapel service on campus. David Sarnoff, president of the Radio Corporation of America, in his address stated that over the years Brown had contributed greatly to the intellectual growth of the nation and had given the nation many great and distinguished leaders. He continued by stating that in the future Brown will also be remembered for another important contribution, The Brown Network, the pioneer campus radio station which had now succeeded in developing an extra-curricular activity, college radio which is a parallel to the college newspaper. He predicted that in the not too distant future college radio would be the greatest training ground for the radio industry. This prediction has come true to an extent even greater that he had predicted at that time. Today thousands of radio and television executives, sports and program announcers and engineers trace their beginnings to college radio stations. Edward Sarnoff, David's son was a Brown student and a member of the Brown Network. After chapel President Wriston invited David Sarnoff to lunch. He accepted with the request that he be able to first visit the original Brown Network in George Abraham's room where, after being introduced by his son, he would broadcast a greeting to the students. The Providence Journal of November 5, 1939 carried a picture of this famous broadcast with the caption *RCA Head Broadcasts over "Rival" Network*. President Wriston and the administration finally realized that this band of students who were climbing over the roofs of Brown's ivy covered buildings, going into the steam tunnels and interviewing students and members of the faculty on campus could contribute greatly to the accomplishments of that staid and conservative university. Indeed the press of the nation agreed that an important new extracurricular activity had been born.

After the visit of David Sarnoff the press of not only Providence but of the entire nation became interested in The Brown Network. As undergraduates, we were in the limelight. The story of our new station was not only important to the future of radio but also was a story of great human interest.

On December 3, 1939 The Providence Journal, in its Sunday edition featured "The Brown Network" in a full page story showing various activities of the station. Photos included the control room at Faunce House, reporters in a boat broadcasting the progress of a dinghy race on the Seekonk River in Providence, a group of students at the Delta Upsilon Fraternity House listening to a "network" program, a broadcast of a football game from the stadium and a one act drama thriller being broadcast with the aid of Pembroke students. Included in this photo was Helen M. Thomas, later to become Mrs. David Borst.

The Christian Science Monitor sent its radio reporter, Albert D. Hughes to Brown and he reported as follows: "You might call this one of the first stories ever obtained by collegiate wired radio. George Abraham, Brown senior, stepped to a panel in his room in Slater Hall on College Hill here, threw a few switches, picked up a hand microphone and in a friendly 'Calling All Cars' manner said, 'Abraham in Slater calling all houses and dormitories.' Back at him through an overhead loudspeaker came nearly a dozen voices, 'Faunce Calling', 'Metcalf Calling' 'Hope Calling' as the various dormitories answered him. 'Listen fellows,' George retorted, 'I've got a reporter and a photographer here and we are coming around to take some pictures so wait until we get there. We would like various groups of you to pose for them. Okay? Swell. We'll be seeing you.' In a few minutes conversation, we had let a fair portion of the university 'in' on our mission there. Was ever a reporter's or photographer's work more easily cut out?"

As news of this new campus "fun station" spread we began to receive cooperation and offers of help from the radio industry and the major networks. The Columbia Broadcasting System allowed a Brown commentator, sent to cover the Brown-Princeton football game at Princeton to give his account to the students back home from the CBS booth. They allowed the Brown Network to connect its rented telephone line to the CBS booth.

As the newspapers of the nation publicized The Brown Network, colleges, mostly in New England, contacted us requesting information regarding construction of stations on their campuses. Dave, always helpful, sent detailed instructions and visited several of the campuses. The second station to commence operation was WES, The Cardinal Network at Wesleyan University at Middletown, Connecticut which made its first broadcast on November 9, 1939 with a daily alarm clock wake-up program from 7:30 a.m. to 8:30 a.m. and the daily "Jive at Five" starting at 5:00 p.m. Soon other programs were added including "The Argus Reporter", "The Symphony Hour" and on Saturday Nights "The Midnight Matinee". President McConaughy of Wesleyan gave official permission to the station for the use of the college heating tunnels for the purpose of running wires to all buildings on campus. Dr. Eaton of the Physics Department was appointed as the faculty technical advisor. A permanent soundproof studio was established in the basement of Clark Hall. A main control panel was installed behind a glass partition which separated the control room from the main studio. The Cardinal Network operated on a frequency of 680 kc. Program Director was Robert Stuart, Technical Director, Archie Doty and Business Manager Richard Coleman.

In 1939 Pembroke College, the women's college of Brown University, was added to the Brown Network by means of leased wire lines to miniature transmitters in Metcalf, West and East Halls. Jean Bruce '40 was in charge of the thirty Pembroke women working on The Brown Network.

The first Brown Network control room, Faunce House, Brown University, 1938.
Left to right, Louis M. Bloch, Jr., unidentified, Ruth Van Dyke, Helen Messinger.

Portable units of the Brown system are taken to all sports events. Here's a broadcast from the football stadium. (Courtesy of The Providence Journal)

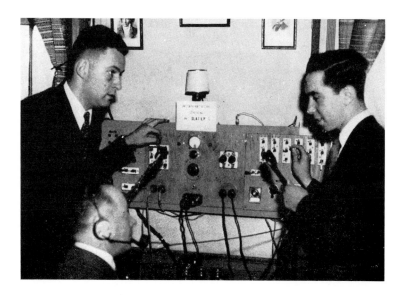

George Abraham describing The Brown Network's operation to David Sarnoff. Left to right: George Abraham, David Sarnoff, Edward Sarnoff.

Remote broadcast of a sailboat race. Left to right: Louis M. Bloch, Jr., Sherwin Drury.

"Jam For Your Breakfast" is the unique title of this early morning Brown Network broadcast. (Courtesy of The Providence Journal)

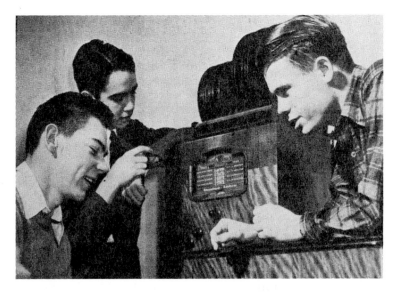

In a dormitory room, students tune in a receiving set to pick up a Brown broadcast. (Courtesy of The Providence Journal)

CHAPTER III

THE INTERCOLLEGIATE
BROADCASTING SYSTEM

News of college radio spread fast as major newspapers and news media throughout the country spread the news of the new extra-curricular activity which was the parallel of college journalism.

It was announced by George Abraham that the first convention of The Intercollegiate Broadcasting System would be held at Brown University February 17th and 18th, 1940. Invited as guest speakers were Dr. Franklin Dunham, Educational Director of NBC and Dr. Loring Andrews, chairman of the program committee of the World-Wide Broadcasting Foundation. Also invited were H. Linus Travers,

Vice-President of The Yankee Network, John Holman, General Manager of WBZ, Boston, Malcolm Parker, supervisor of station WEAN, Providence and James A. Williams, District Manager of the New England Telephone Company. President Wriston of Brown welcomed the delegates. George placed the Brown athletic director, who had pulled out the tubes of the radios in Faunce House during a football broadcast so the students would go to the game, between two important radio executives and directly across from the Brown Network Board at the dinner. Also the Assistant Dean who had caused us no end of trouble was placed in the midst of radio executives who, at the dinner, constantly praised college radio. George Abraham, founder of the Brown Network and the father of college radio was elected Chairman of the Intercollegiate Broadcasting System, and assisting him were Peter Thorpe, Advertising Manager, David Borst, Technical Manager, and Joseph Parnicky, Program Manager. I was elected Business Manager and it was my job to obtain advertising for the college radio stations. All were volunteer jobs as IBS had no income of any kind. A constitution and by-laws were adopted. Once broadcasting systems were established at the other colleges, the IBS was to act as a medium through which programs and ideas were to be exchanged as well as an agency to obtain national advertising. No budget or any other monetary consideration was even discussed at this first convention. It was our conviction that things would work out somehow. We never had any money so why worry about it. As I learned, personal sacrifices had to be made but at that exciting moment I did not realize that I would be making most of them.

Colleges which sent delegates to this first IBS Convention were Brown, Columbia, Cornell, Dartmouth, Harvard, M.I.T., Pembroke, Rhode Island State, University of Connecticut, University of New Hampshire, Wesleyan and Williams. Other colleges had planned to attend but were unable to get to Providence because of the greatest snow storm of the year which closed the airport and made travel difficult.

On March 18, 1940 the Williams College station using the call letters WMS commenced broadcasting. The WMS studio sent its programs by wires to transmitters located in five dormitories. The dormitory transmitters were connected with the heating pipes in each building. This allowed any regular receiving set to receive the station within an area of 100 feet from a heating pipe. Initial programs consisted of music, faculty quizzes, sports events, news and drama.

The students at Kent State University, Kent, Ohio were planning a college station, so during my spring vacation, I visited the Kent campus. On arrival on April 3, 1940 I was escorted to Dr. Leebrick's office and as president of the University, he welcomed me to the campus and affirmed his great interest in college radio. The Radio Workshop with 66 members had been organized on February 27, 1940 and planning for studios and the control room had just begun. I was given a tour of the proposed area for the new station. As I was going through the hall a student rushed up to me and gave me a copy of *The Kent Stater*, the daily newspaper which had just been released. The headline read IBS MAN ARRIVES TO PLAN PROPOSED RADIO STATION.

Studios for The Radio Workshop were completed in the Summer of 1940 and students originated broadcasts for station WADC, Akron, Ohio. Due to the war the wired wireless station did not comence operation until 1948.

Through the efforts of Professor Walter Clarke, professor of speech and director of the Kent State Radio Workshop, and Professor W. Turner Stump, director of the School of Speech, the first university FM station operating with 10 watts was approved.

In 1952 Dr. John C. Weiser took over major control of the FM station. He served as General Manager until 1971. When he returned to the teaching of the radio and television classes, John Perry became General Manager. Today, mainly through the efforts of Dr. John Weiser, Kent boasts one of the finest radio and television complexes in the nation. This complex includes WKSU-FM, a 50,000 watt National Public Radio station which uses a satellite to rebroadcast programs of any of the 220 National Public Radio stations; WKSR, a campus confined station covering the University dormitories with a potential audience of 7,000 students; a closed circuit television station confined to the campus and the finest radio and television studios for instruction.

On May 4, 1940, the Intercollegiate Broadcasting System, with the cooperation of station WRUL, The World Wide Broadcasting Foundation, made its first intercollegiate broadcast. This was a series of five weekly broadcasts featuring Brown University. The first program marked the completion of the Colonial reconstruction of University Hall, Brown's original "College Edifice" of 1770. These programs were carried by campus stations at Harvard, Williams, Wesleyan, University of Connecticut and Rhode Island State. Arranging this series of programs had brought problems from our major critic, the Dean. He had never approved of The Brown Network, or of the students climbing over his ivy covered buildings. To him college radio seemed to be an outrageous development and now, without his permission, this unruly group had planned a network broadcast which would cover New England. He notified George that the programs must be cancelled and that if we did not agree to his demand he would inform the Vice President and even the President of the University. George did not agree to terminate the broadcasts and the Dean promptly called Vice President Adams who informed him that he was pleased that Brown was to be honored by originating the first college network broadcast. Furthermore he stated that he was a featured speaker on the closing program and that President Wriston was the featured speaker on the opening broadcast.

We had one last tiff with the Dean who by this time should have learned his lesson. The Dean had grievances about The Brown Network and summoned George and me to appear in his office the following week. In the meantime George sent out a national news release regarding the rapid development of college radio and listed the colleges who were constructing stations modeled after The Brown Network. When we arrived at the Dean's office for the meeting he stated that we had no business climbing over the buildings, and stringing our lines through the heating tunnels. He then opened his desk drawer which revealed hundreds of press clippings received from almost every state describing the great contribution which Brown had made by developing a new extracurricular activity which was in step with the times. Also the press praised Brown for permitting an activity which was certain to become a most important training ground for the radio industry. The Dean's only comment was, "So What Can I Say". After that incident the Dean became a good friend and we had no further problems.

Final examinations were approaching and we hoped that graduation would follow but the press of the nation continued to hound us. George had the misfortune of breaking his leg and was confined to the college infirmary. Not to allow such a minor occurance to hinder him in any way, George called a news conference in his hospital room. It was well attended by the press. George was and is a master of organization and with absolutely no money he had succeeded in making this new phenomenon, college radio, the talk of colleges from coast to coast.

Faunce House, Brown University, the home of the first IBS convention.

CHAPTER IV

GRADUATION

My senior year at Brown was drawing to a close. The country was in a depression and war clouds were moving on the horizon. Graduation from Brown was a high point in my life, as it was in the lives of my parents. The three days of graduation festivities were glorious. It was the custom at Brown to have an evening campus dance on the green in front of Faunce House, the social center of the University. Paper lanterns lighted the dance floor and tables surrounded the dancing area, each assigned to a fraternity or dormitory. On the north side of the Green stood Hope College and University Hall, the original buildings of the University. During the Revolution Lafayette had quartered his troops in University Hall and in June, 1940 the building had been beautifully redecorated inside, although the outside looked the same as in the early prints of the University. Hope College, a campus dormitory, looked internally and externally just as it looked in the early 1820's or so it seemed to us. The steps were grooved where Brown men had climbed and descended them for over a century.

The commencement march began on the college green where the lines for the march formed. The procession was headed by the President of the University followed by the trustees in their pork pie hats and the professors all garbed in their flowing academic robes. The High Sheriff of Providence, dressed in a high silk hat, a blue sash and sword, still participated in the commencement ceremony just as did his predecessor in 1791 when the Rhode Island General Assembly moved to curtail raucus disorderliness at the "Meeting House".

Following in the procession were the alumni classes starting with the earliest and then our graduating class, the class of 1940. To the music of a marching band this procession wound its way down "College Hill" to the First Baptist Meeting House. The Baptist Church was founded in Providence in 1638 and "The Meeting House" which stands today was built in 1775 at the outbreak of the Revolution. The church interior was a sight to behold and a chill went up my spine as I entered with my class. The architecture was plain colonial with the most beautiful chandelier I had ever seen. It was made of Waterford glass crystal and was given to the church in 1792 in memory of Nicholas Brown by Hope Brown, his daughter, and was lighted for the first time on the evening of Miss Brown's marriage to Thomas Poynton Ives in 1792. It burned candles until 1884, then gas and in 1914, electricity. The organ with many of its original pipes intact was installed in 1834, the gift of Nicholas Brown 2nd. The big bell in the steeple, made in England and recast several times, still is rung every Sunday and at times of special import. The complete rehabilitation of the Meeting House in 1957-58 was made possible by a gift from John D. Rockefeller, Jr.

As we entered "The Meeting House" for our graduation, we were following Brown graduates, many of whom had put their stamp on America. Brown never boasted a President but had four secretaries of state and many other distinguished Americans, and some not so distinguished.

After the graduation ceremony, my former roommate, Jerry Strauss and I and our parents went to a charming country inn for dinner.

So ended my years at Brown. Looking back now, I was an average student, an economics major who didn't believe in the theory of economics as it was then presented. I enjoyed Greek Civilization, Music Appreciation and other courses but I realize that my greatest interest was The Brown Network and The Intercollegiate Broadcasting System. I think that I learned more from George Abraham than I did from all of my college courses. George was and is an organizational genius who lets no barrier stop him even though the accomplishment of the goal might seem impossible.

With our college years at an end, George & Dave took jobs in the east as electrical engineers and I returned home to Cleveland with my parents. I had been elected Business Manager of The Intercollegiate Broadcasting System and was given the job of setting up a national office and of obtaining national advertising for the member IBS stations. No compensation or expense accounts were discussed because we had no money. I managed to stay home for one month and then decided that it was my obligation to assume my new responsibilities. Before college I had delivered newspapers for many years and had built up a small bank account. I withdrew almost all of my savings and soon was on a train for New York City.

The Kent Stater

Z 568   Kent State University, Kent, Ohio, Wednesday, April 3, 1940   1

IBS Man Arrives To Plan Proposed Radio Station

Possibilities for a radio station on the campus, were further enhanced today with the arrival of Louis M. Bloch Jr., business manager of the Intercollegiate Broadcasting System, who is paying the university a visit of "goodwill."

Bloch, student at Brown university, and a resident of Cleveland, arrived this morning with the intentions of planning the proposed radio facilities and laying out the groundwork in theory.

IBS has been broadcasting all university programs over their network of eastern schools for the past four years. Tentative plans here await sanction from the administration for money which would appropriate lines and prepare the workshop for broadcasting schedule.

Bob Stephan, radio editor of the Cleveland Plain Dealer, gave Kent mention in his column yesterday with "am extremely interested in the work J. F. Redmond and staff is doing and would like to know more about their plans."

CHAPTER V

NEW YORK — 1940

My arrival in New York City in August of 1940 was a high point in my life. I checked into the YMCA and called my uncle, a writer and publisher of crossword puzzle magazines. He was amazed at my brashness at arriving in New York without a paying job and with no immediate prospect of an income, however, he did everything he could to help me. It was good to have relatives in New York especially when I had to operate on such a tight budget, having to watch my expenditures including my rent and meals. My aunt and uncle invited me to dinner every week or so and that lifted my spirits. My room rent at the YMCA was only $5.00 per week and it was possible to get a good lunch for 50¢. A 25¢ breakfast was usual. The depression had been with the nation and war had broken out in Europe. I registered for the draft but the Army turned me down because I had had infantile paralysis when I was six years old and my shoulder had been slightly affected. That early bout with infantile paralysis enabled me to continue my efforts for IBS.

My uncle lent me a directory of advertising agencies and I set about the job of attempting to interest national accounts in advertising over the college radio stations. At first I met with no success and was laughed out of most of the advertising agencies which I contacted. I was a kid of 22 years representing "The Gaspipe Network" as the Intercollegiate Broadcasting System was affectionately called.

Luck however changed when George Goldberg, the son of Rube Goldberg, the famous cartoonist, returned home to New York. He was interested in acquiring advertising for his college station, WMS, Williams College. Together we approached Marschalk & Pratt, an advertising agency which represented ESSO, The Standard Oil Company of New Jersey. George wanted ESSO to sponsor the Williams-Army football game for that coming Fall at Williams but ESSO wanted continuity and finally signed for five ESSO 5 minute newscasts per week at Williams and added the same for Brown. The broadcasts were to be prepared and delivered by students. Finally the ice had been broken and Tide Magazine featured an article entitled *Short Cheer ESSO* describing how a tall, curly-headed radio bug by the name of George Abraham started these gas pipe stations at Brown and now they were worthy of the dollars of national advertisers. The article also announced that I would open Manhattan offices to solicit national advertising for the twelve college IBS member stations whose administrations permitted paid advertising. Radio Daily, The National Daily Newspaper of Radio and Television featured an article entitled, *2 Watt College Web Gets a National Sponsor.* Their article opened as follows, "Believed to be the smallest radio stations ever signed by one of the largest national advertisers, The Standard Oil Company of New Jersey, announced over the weekend that it had contracted for a series of daily newscasts over WMS, a broadcasting station serving dormitories and fraternity houses at Williams College, Williamstown, Mass. and The Brown Network covering Brown University, Providence, R.I. Both of these stations, operated with two watts by and for students, transmit news of campus happenings to radio sets within college buildings. The collegiate Esso Reporter will broadcast nightly using items edited and gathered by the student staff. This new program series, inaugurated Sept. 23, 1940, was the first program to be sponsored by other than local advertisers."

Advertising Age, The National Newspaper of Advertising featured a story with the headline *Esso Reporter to Do Stuff on Collegiate Net.* Dated Sept. 11, 1940 the story reported as follows: "The Esso Reporter, whose newscasts are heard nightly from Maine to Louisiana and who makes a weekly television appearance, will go collegiate September 23rd when a new series of nightly news programs will be inaugurated via the 'Gas Pipe Network'. This network including Station WMS of Williams College at Williamstown, Mass. and the 'Brown Network' of Brown University at Providence R.I. gained its inaccurate name from the fact that transmission wires linking a two watt transmitter with receiving sets in dormitories and fraternity houses on each campus are strung through underground pipe conduits connecting the buildings. Both stations are undergraduate ventures operating unfettered by Federal Communication Commission rules." Broadcasting Magazine headlined the story *"Gas Pipe Networks Sign ESSO, Wired Hookups at Williams and Brown University Get News Sponsor Outside FCC Sphere."* The Esso Reporter featured local campus news prepared and delivered by student newscasters. The news of our success exploded in the press across the country and my reputation as a radio executive was established, even though I had no income and no office.

Up to this point I was paying for all New York City IBS activities out of my own pocket and my savings were starting to shrink. The $5.00 per week rent at the YMCA was quite reasonable and I was spending about $15.00 per week for food, transportation and other expenses. Prices in those days were a far cry from today's prices. The subway was 5¢ and movies were 25¢.

After signing the ESSO contract, I celebrated with a dinner at the Fleur De Lis restaurant on 79th street near Broadway. The dinner consisted of six trays of hors d'ouvres and a full French dinner, the like of which today would cost $15.00 or more, but in September of 1940 the price was $1.00. Dessert was extra but the dinner was so filling that I did not take it.

One day when I was especially low on cash I went into a New York bar and purchased a beer for 10¢. Trays of bread, ham, beef and cheese were available to all bar customers at no charge so I had a fine lunch with beer for 10¢. I was impressed with a restaurant in Grand Central Station which allowed patrons to select anything desired, each item being priced, allowing the customer to total the price and pay according to his total on leaving the restaurant. It was found that even in New York 98% of the people were honest. This policy of trusting people enabled the restaurant to hire fewer employees and so maintain low prices. They did a big business.

I soon realized that at a $20.00 per week rate of expenditure I could not stay in New York unless I made other arrangements to obtain an income. I had just about gone through my savings. IBS had little money and was operated on "nervous energy" and the personal contributions of its few dedicated founders. At the suggestion of my uncle, I contacted a radio station representative on Madison Avenue to represent the college radio stations and pay me a salary for handling the account. The company contacted was Weed and Company directed by Joseph Weed. He was intrigued with the idea but felt that an association with the "Gas Pipe Networks" would make him the laughing stock of New York. Swearing me to secrecy, he agreed to set up a station representative agency to be known as "Intercollegiate Broadcasting Station Representatives". This IBS representation by him has not been revealed until now, forty years later. Through the interest of Joe Weed, IBS was able to establish itself in New York. Weed rented an office at Fifth Avenue and 42nd Street diagonally across from the New York Public Library. The first office consisted of a small room sublet from a Mr. Blood. The other tenant in the office was Julius Whitmark, of "Tin Pan Alley" fame, one of the founders of the Whitmark Publishing Company, the great music publisher. At that time Julius Whitmark was 88 years old and was writing his memoirs. My salary as manager of IBSR, as our agency became known, was $25.00 per week plus expenses and I was instructed to report each Friday to Mr. Weed at his Madison Avenue office to make a status report and pick up my check. In that way Mr. Weed, a successful businessman, kept in constant touch with IBS and college radio through me from Sept., 1940, to February, 1946 when I resigned.

Finally IBS had an office in New York, as Joe Weed was allowing us to use the IBSR office as IBS headquarters. I still continued to be Business Manager of IBS but all of my income was paid by Joe Weed.

With the formal introduction into the radio field with a New York office and an agency to represent us, the industry pricked up its ears, but we were still the laughing stock of the radio industry.

The second IBS convention took place at Columbia University on December 27th & 28th, 1940. Colleges attending were Columbia, Cornell, Hamilton, Harvard, Haverford, Maryland, Ohio University, Rhode Island State, Swarthmore, Wesleyan and Yale. Joe Weed had approved the hiring of a secretary for the Intercollegiate Broadcasting Station Representatives. Naomi Fine was selected at a salary of $20.00 per week. She attended the convention, with George, Dave and myself representing IBS.

In January, 1941, Variety Magazine sent a reporter to Williams College to investigate the new phenomenon of campus radio. The reporter spent two days at Williams and on January 22, 1941, Variety broke a feature story with the bold headlines *College Boy Fun Stations Actually Draw National Advertising Attention. Esso, Beechnut, Stanford Ink, Biltmore Hotel on Williams College outlet.* I signed those accounts and Joe Weed was pleased.

A rate card was developed by IBS listing colleges by enrollment. Group A included colleges with 5,000 students or more; Group B, 2,500-5,000; Group C, 1,000-2,500 and Group D, colleges with less than 1,000 students. Rates ranged from $30.00 per half hour on Class A stations to $12.00 on Class D stations. Fifteen and five minute segments were proportionally less. One minute spots ranged from $5.00 on Class A stations to $2.00 on Class D stations. Of the amount received from advertising, 15% went to the advertising agency representing the account, 15% to Intercollegiate Broadcasting Station Representatives, 20% to IBS and the remaining 50% to the college station. Finally both IBS and the individual college stations had a reliable source of income.

Success stories of college radio obtaining national, advertising now became the news of the radio world. Broadcasting Magazine coined a new phrase, "One Lungers", in their feature article about college radio in their issue of March 17, 1941. New "One Lungers", they reported now include Cornell, Princeton and Columbia in addition to the existing stations at Brown, Williams, Wesleyan and Rhole Island State. Broadcasting Magazine also reported success stories for local advertisers. At Princeton, "a local jeweler who used two half hour programs for a sale promotion sold 45 pieces of jewelry to the students as a result of the advertising." "At Wesleyan, a local record dealer provided the station with swing and classical recordings and bought time to promote the sale of the records. The sale of these recordings to the students soared."

The North American Newspaper Alliance sent out a national news release with the headline *Gaspipes To Vie With Big Chains*. A cartoon showed a student listening to his radio which was tied to his radiator. The article predicted that the commercial networks would have serious competition if "gaspipes" continued to spring up on college campuses across the nation. For the first time the prediction was made that eventually college stations would use FM as their mode of broadcasting as well as carrier current. That prediction has come true.

Business for college radio was good and the advertising agencies welcomed me as I made contacts. As IBS started to receive income from advertising revenue and dues, IBS decided to officially merge its office with IBSR, the IBS station representative of which I was Manager. I was simultaneously Business Manager of IBS, in which job I served without salary, receiving my sole $25.00 per week income from IBSR. IBS and IBSR sharing expenses rented a full office in the same building, 507 Fifth Avenue diagonally across from The New York Public Library. Lawrence Lader, a graduate of The Crimson Network at Harvard, and Sonia Jane Brown, a Pembroke graduate, assumed many of the program and administrative responsibilities of the Intercollegiate Broadcasting System.

With the cooperation of Larry Lader and Jean McGinnis, we developed a brochure entitled *From Princeton to Stanford, IBS sells the Colleges.* This brochure told the story of a day in the life of Toby Green, a hypothetical Williams student and outlined his buying habits and the market which he and his fellow students represented. It described the influence of the college student on the young adult market. Toby's radio listening habits were described. That information was gleaned from a survey which I conducted on the campuses of Princeton, Wesleyan, Union, Connecticut, Williams, Cornell and Columbia. This entire brochure is reprinted in the appendix.

Two major accounts were signed, Readers Digest and Camel Cigarettes. Readers Digest used spot announcements and R. J. Reynolds Tobacco Company sponsored "The Camel Campus Caravan", a fifteen minute, three times a week, popular music program.

In a feature article, The Chicago Sun Times referred to the college stations as operating on "mouse power". College stations began to spring up throughout the country and were being constructed, in some cases, at a cost of less than $200.00. The Chicago Sun Times reported that a listening survey had been conducted at one of the "mouse power" stations, The Brown Network. The survey was described as follows: "Students who happened to be listening to their radios at Brown University, Providence, R. I. recently heard the announcer say, 'At a given signal will all who are listening, raise your windows and holler.' The signal was given and windows flew open all over the campus. The roar that followed could be heard for blocks. Thus was completed the quickest, most direct and certainly the most vociferous radio listener survey in the history of broadcasting."

At the age of twenty-three I was Business Manager of a Broadcasting System and Manager of a radio station representation organization. Joe Weed increased my salary to $35.00 per week. I had also become the laughing stock of the radio industry and was the butt of many jokes to which I did not at all object as they created publicity for college radio. It was at that time that I joined "The Broadcasters Bull Session", a group that met each Friday for lunch at Stouffer's Restaurant on Fifth Avenue.

The group consisted of people from the networks, advertising agencies and media representatives. It was at these luncheons that I made valuable contacts, some of which resulted in contracts for time on the college stations. The interest in the "Broadcasters Bull Session" was tremendous and they soon needed much larger seating capacity for their luncheons. The group changed its name to the Radio Executives Club of New York and when I left New York in 1946, they were renting the ballroom of the Pennsylvania Hotel for their meetings.

On May 24, 1941 college radio received its greatest publicity with a feature article in The Saturday Evening Post. The article entitled *Radiator Pipe Broadcasters* by Eric Barnouw told the story of a college dean in his home on the edge of the campus tuning his radio to the late night news when he suddenly found a very loud radio station which he never knew about, playing Sibelius' Second Symphony. The dean was sure that the station was nearby but he could not find it listed in the newspaper. When the music ended a voice declared, "This is the college broadcasting system station of_____college." The alarmed dean realized that the broadcast originated from his own campus. His cause for alarm was his knowledge that it was illegal, under the rules of the Federal Communications Commission to broadcast without a license. He also knew that the FCC would hardly give a license to students to operate radio station in the standard band. Immediately the dean called the campus police and instructed them to locate the transmitter with the aid of a portable radio. The students were found and ordered to appear the following day at the dean's office. The next day when the students arrived at the dean's office, they told an incredible story.

They described how the radiator pipes of the college could be turned into a transmitting antenna and how the college electrical wires, gas and water pipes could be used for the same purpose, but because low power signals do not radiate for a great distance from the pipes and wires, the operation was not considered a broadcasting station by the FCC which was not interested in gadgets such as wireless phonographs which broadcast across a room or only a short distance. Generally a transmitter broadcasting less than 200 feet from electrical lines or pipes was safe from any Federal interference. The Saturday Evening Post article referred to the college stations as "flea power" stations, a term that was repeated in the nation's press.

This publicity alerted students throughout the United States and foreign countries to this new extra curricular activity, college radio, which was the parallel of college journalism. Letters began pouring in from students everywhere and Dave Borst was hard pressed to answer all of the technical questions posed. College authorities also realized that something important was happening on their campuses, that a new age had arrived. As the college newspaper was a great training ground for journalism, so now, the college radio station was destined to become the most important training ground for the radio and eventually the television industry. As of today, 1980, many of the top executives of both the radio and television industry got their start working on college radio stations. Over one and a half million individuals have been members of college stations since 1939. It was not unusual to have two to three hundred students at a single college involved in some way with the college station. This great number developed because of the many students needed to construct and maintain the studios, transmitter, lines, the sports, the news and the program department. Most stations also had a business department to handle finances and obtain advertising. In some cases stations cooperated with the college band, orchestra, and dramatic organizations in airing their programs.

As a result of the great amount of publicity college radio was getting, The Federal Communications Commission issued the following release:

"Development of so-called 'intercollegiate broadcasting systems' has prompted numerous inquiries to the Federal Communications Commission about this newcomer in the field of low power radio frequency devices. In response to one such inquiry from Los Angeles, the Commission replies in part:

'In the intercollegiate broadcasting systems communication is effected not by the transmission of radio waves through space but by the transmission of radio frequency currents via wire lines. Radiation of energy from the lines capable of causing interference is prevented by proper shielding of the lines in metal conduit. You may obtain further information of such systems from Mr. David W. Borst, Technical Manager of the Intercollegiate Broadcasting System, 13 State Street, Schnectady, N.Y.

'Preliminary investigations have indicated that these intercollegiate systems are well engineered and supervised. No interference has been reported as a result of their use. The Commission has therefore not promulgated any rule governing their operation.

'This type of system, however, if used on open lines or if

improperly designed, is capable of causing very serious interference. The Commission is therefore making a study with a view to the need for regulations in the case of extension of this principle of communication into other fields."

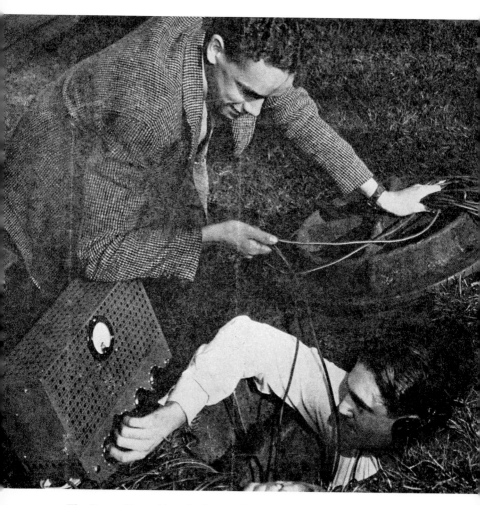

The Brown Network's web of transmission lines follow the university's steam tunnels around the campus. Here, Trouble Shooters Linford S. MacDonald (in manhole), Pittsburgh, Pa., and Ralph E. Waters, test the wires to make sure that all is well. (Courtesy of The Providence Journal)

CHAPTER VI

THE OLD COLLEGE TRY

"There aren't many people who would undertake a project to write, direct and produce a radio program in a foreign language, for instance, Portuguese. Even skilled program directors would hesitate at a project like that but the college kids will try anything." So reported the New York Times of September 7, 1941. Believing that students everywhere have a great deal in common, IBS during the Summer of 1941 formed a Pan-American Department. Lawrence Lader of The Harvard Crimson Network was made Director. A series of programs for transmission between IBS and South American universities was developed. IBS was to send out a fifteen minute program in Spanish or Portuguese directed to particular universities in South America and the plan was for return broadcasts in English from these universities. NBC offered their shortwave facilities. The plan was discussed with a Brazilian student living at International House in New York three blocks from the Columbia campus who was returning to Brazil. The student agreed to develop the return broadcast. Unfortunately all efforts to contact the student, who had left with the best promises, failed. Later one of his friends told me that he probably went to the Amazon for a vacation. At any rate, on August 15, 1941, our program to the University of Sao Paulo from various IBS college stations went off on schedule. We learned the hard way that "manana" was a way of life in South American as no return broadcast was received.

With such a fiasco behind them, The Pan American Department made a second try. This time the nation of Columbia was selected and so to take no chances for a failure of a second exchange, the president of the selected university who was visiting New York was asked to arrrange for the return broadcast. The time and outline of the broadcast were all planned in New York before he left for Columbia. Everything seemed to be fine until a few days before the broadcast when the following letter was received from the Columbian university, "Yes, we would like to send you a program in English, but it hasn't worked out yet, however it will be ready sometime soon". IBS, however, sent its program in Spanish to Columbia where it was broadcast on the Federal Columbian station. This second program was a tribute to medical and engineering students. They represented the two most needed professions in Columbia. The writing of these programs was done in English and translated into the languages of the countries, either Portuguese or Spanish by South American students living at International House. These students also participated as the actors and as announcers under the expert supervision of the NBC announcers. Other exchange programs to Mexico and Peru were planned but because of "manana" they were finally cancelled.

Although the Pan American Department did not achieve all of its goals, a start was made in the exchange of student broadcasts between the Americas. In commendation for his efforts Lawrence Lader received letters of congratulation from J. W. Studebaker, Commissioner of the U.S. Office of Education, John M. Begg of the Cultural Relations Division of the Department of State and Eleanor Roosevelt.

Halloween at International House. The British group enacts the bombing of London.

CHAPTER VII

THE FIRST YEARS OF THE WAR

In August of 1941, I had moved to International House, a twelve story residence for foreign students, located at 123rd Street and Riverside Drive. Men and women lived on alternate floors and, although I was not a student, I was permitted to rent a room there. The price of $5.50 a week for a room was quite a sum for me in those days but meals in the cafeteria were inexpensive and the subway fare to the IBS office was only a nickel. Also the Fifth Avenue double deck bus ran on Riverside Drive to Fifth Avenue and I could get off in front of my office. The fare for the bus was ten cents, so I usually took the subway.

Burl Ives had moved into International House several years before my arrival. In those days Burl had little money and had to make ends meet in the best possible way, as we all did. A story had been making the rounds of International House regarding Burl Ives' first visits to International House. It seems that every evening about nine o'clock a long black object passed the windows of the rooms on the girl's floors on its way to the roof. Some of the girls were terrified until the mystery was solved. It seems that Burl Ives was

51

pulling his sleeping bag to the roof each night and sleeping there. I never knew of this story from first hand experience and it may or may not be true. By the time I arrived at International House, Burl Ives was a distinguished resident. His rise to fame had begun with his network radio programs of folk music. He was always a familiar sight at International House and participated in many of its activities. I remember especially the picnic for House members which I attended when Burl Ives sang his glorious folk songs by the light of the camp fire. He closed his program with his ever favorite "The Blue Tailed Fly".

At International House it was the tradition to have Sunday night suppers for which there was no charge. I hardly missed a supper as it was a stirring event with always a good speaker, and, of course, it was free. Students from over sixty countries lived at International House. Some of my best friends were from Afghanistan, India, Iran, Brazil, France, Britain etc. The French, British, Belgian, Dutch, Norwegian and those from other western european countries were much distressed because of the war. We Americans realized that it wouldn't be long before we also would be involved.

Each Halloween a big carnival was featured at International House. Each nationality put on its own show. In October of 1941 I decided to join the British group and played the part of a London bobbie during a London blitz. We had actual recordings of an air raid with the bombs dropping and the sirens screeching. Ours was a dramatic production which shook the building. One of the Nigerian students, a prince from the royal family was anxious to participate in the Halloween carnival but the Nigerian group had no organized activity. He developed his own activity by persuading the carnival director to allow him to erect a high chair at the International House entrance where he could sit in his princely robes and his gold cane with a sign "INFORMATION". He added a flavor to the festivities from the very moment a visitor entered the doors of International House.

International House, as I mentioned, is a twelve story building with six floors reserved for men and six floors for women. The men's floors were even numbered while the women had the odd numbered floors. Elevators at each end of the building served respectively, men and women. The downstairs consisted of common rooms including an auditorium, music room, gymnasium, meeting rooms and a large lobby running almost the entire length of the building. In the basement there was a cafeteria, the Waffle Wing which was a snack shop, a barber shop and a stationery store. The administration staff was rigid, and I must say effective in keeping the living quarters separate.

It was into this atmosphere that we introduced the clandestine radio station WOLF, "The Voice of International House". WOLF was a two watt station which was fed into the electric light system of the building. Programs were at odd hours and consisted of records, skits and talk shows featuring both men and women. The location of the station was secret, and the staff was alarmed and helpless. They were convinced that the programs with mixed sexes must be originating from one of the rooms. They made a thorough search of the common rooms and then secretly made a search of the entire building without finding a trace of the station. The broadcasts continued and the Director was frantic. He was faced with a breakdown of the long and successful separation of the living quarters of the sexes. How could he explain this to his board which supplied all of his financial support? I guess we were just pranksters at heart looking for a good prank but seeing the genuine distress we were causing, we decided to suspend operations. We announced over station WOLF that we were terminating our broadcasts and thanked everyone for being such a loyal audience. Everyone but the staff got a kick out of our experiment but, until this day, no one except our station members knew how these broadcasts were accomplished. Our chief engineer had built the transmitter in an old suitcase which he kept under his bed and all broadcasts originated from his room. To obtain the women's participation in the programs, he had simply lowered a microphone to the women's floor below.

The Columbia University station was getting its start and I spent a good deal of time at the station which was located in a small building situated between Hartley and Hamilton Halls on the Columbia campus. The station's location was ideal, being at the heart of the campus. CURC, The Columbia University Radio Club, was the brainchild of Bill Hutchins '39, who by February 24, 1941 when the station commenced operation, was the assistant to Major Armstrong of Columbia University, the inventor of FM. Starting from the initial broadcasts of recordings and new broadcasts, cooperation was obtained from The Spectator, the campus newspaper; The Debate Council; The drama group and Columbia's civilian defense program. Soon there were broadcasts of campus debates, political and strategical analyses, dramatic thrillers and public addresses. Remote broadcasts were made of pre-football game pep rallies. It wasn't long until CURC had extended a wire line under Broadway to tie in Barnard, Columbia's sister college. Soon the Barnard girls joined the Columbia boys in the broadcasting schedules. The CURC president was Carl Carlson '43 and the program director was Lincoln Diamant '43. The enthusiasm of the Columbia and Barnard students for the station was tremendous. On March 9, 1942 CURC inaugurated a schedule of eighteen hours a day of broadcasting and the station became the center of the social life of the university. CURC made arrangements with W71NY and W2XMN, FM stations to rebroadcast some of their programs. There were many program innovations on CURC including a series of weekly French programs which consisted of light, humorous dramatic shows, composed in easy French and spoken slowly for students taking French courses. The idea was met with enthusiasm by the French faculty and from Le Cercle Lafayette, the French society on campus. Also formed, under the direction of Bill Grauer '43, the station's coordinator of popular music, was an informal "jamsession" group consisting of jazz musicians who participated in unrehearsed jam sessions featuring the early Chicago and Dixieland syncopation.

On Sunday, December 7, 1941 George Abraham had called an executive committee meeting of The Intercollegiate Broadcasting System to meet at his home, a four story brownstone off Central Park. After an excellent lunch we began our meeting. Soon an urgent call was heard from the next room. We all rushed in and found George's sister listening to the radio. The announcer was describing the bombing of Pearl Harbor. George called in, "the news will wait, you can read about it in the newspaper, let's get on with the meeting". The next morning at the IBS office we all pledged to do all we could for the war effort.

As the war progressed, we found, to our surprise, that our talents were in great demand by the Government. We were only a bunch of college graduates in our early twenties but through our experience in developing wired radio frequency, using low power in our broadcasting stations and by striving to keep our signals from radiating beyond the college campuses, we had developed just what the government needed. The Department of Civilian Defense was attempting to develop a system which would alert citizens to bombing raids by radio without allowing the radio station to be a homing beacon for enemy bombers. A major electronics company had developed a special receiver which would accomplish this by picking up stations from many areas but the cost was astronomical. George suggested that he could protect major cities by low power broadcasting using the electric power lines as conductors. He proved this himself by successfully covering a portion of Washington, D.C. During the war, we advised the War Department on the installation of carrier current stations at military bases. While many companies made huge profits from the war, we never charged one cent for our services.

The third convention of The Intercollegiate Broadcasting System, hosted by the Columbia University Radio Club, was held December 29th and 30th, 1941 at International House in New York. Representatives from 20 colleges attended. James Sondheim, Columbia '42 was chairman of the convention.

Eric Barnouw, instructor of radio at Columbia opened the assembly. Sterling Fisher, winner of the Annual Conference of School Broadcasting's award for the best educational broadcast of 1941, followed Mr. Barnouw.

Earl McGill, Dramatic Director of CBS, Walter Craig, Program Director of WMCA, Mrs. Elliott Sanger, Program Director of WQXR, and John Carroll, Public Relations Counsel of CBS took part in informal panel discussions.

On April 7, 1942, The Columbia Spectator published the following article:

CURC's BRONX ZOO EXPEDITION
WILL BE AIRED TONIGHT AT 9:30
Two weeks age CURC went Lion hunting at the Bronx Zoo; tonight it'll let its listeners in on what happened on that much publicized expedition when it broadcasts a fifteen minute recording of the event as it occurred at the Bronx animal emporium.
Scheduled for 9:30 p.m. to 9:45 p.m., the broadcast is a tale of two hours of wild antics performed by the station's members in an effort to evoke a roar from the Bronx felines suitable for a broadcast signature. In their attempt to elicit the

desired sound from the uncooperative cats, the CURC staff did everything from dancing in a leopard skin to masquerading as a lion. A description of these goings-on by Lincoln Diamant '43 and Eddy Costikyan '45 together with a roar finally submitted from an unexpected source comprise the broadcast.

Tonight's recording is a composite of three separate disks made at the zoo. Performing a delicate operation, Engineering Director Martin Scheiner '43 E selected the best parts of the original records and welded them into a compact fifteen minute waxing.

The entire stunt was organized by Eugene Saerchinger '43, CURC publicity director. Scheiner directed the engineering angle while Diamant composed the program.

Starting tomorrow, the lion's roar will be used as the station's sign-on and sign-off and occasionally as a break. Columbia's "Voice of the Roaring Lion" has finally found a voice".

Now CURC had its opening and closing signature and it wasn't long before the roaring lion would be put to an even more important use. In cooperation with the Department of Civilian Defense, we had been designing, a plan to alert civilians to air raids, without broadcasting over a standard radio station which would act as a beacon to approaching bombers. Our plan was to put on a mock air raid drill in the dormitories of Columbia University. We asked the men in the Columbia dorms to leave their radios on all night at a clear designated frequency. If they heard an alarm, they were to go downstairs, check their names and return to bed. At exactly 2:00 a.m. on a Tuesday night, after calling the press and the college administration, the alarm was sounded. It was the "Voice of the Roaring Lion" which reverberated through the dormitories. The announcement was made "This is a practice air raid drill, follow instructions, leave the building, check in and return to bed." The drill was a great success and the press had a field day.

My years at International House brought me in contact with students from every part of the world and broadened my horizon. I became color blind and any prejudice which I might have had vanished. Living at International House were Moslems, Christians, Parsis, Jews, Agnostics and many more. All continents were represented with many Latin Americans, Europeans, Indians, Africans, Asians. About 20% of the residents were North Americans. English was understood and spoken by everyone. We went hiking, camping and dining together. The Chinese restaurant on Broadway at 123rd Street was a favorite hangout for us. The Chinese students especially enjoyed the original Chinese cooking while we preferred the standard American style Chinese food.

When war was declared, all Japanese were rounded up and taken to internment camps. Among those taken from International House was a Japanese girl who was a friend to all. That was a really sad day.

My Latin American friends knew that I was involved in college radio. One day when we were discussing radio broadcasting, they confided in me that broadcasts which our State Department was sending to South and Central America were doing a great deal of damage to our relations. When I asked why this was so, they stated that the State Department broadcasts were talking down to the Latin Americans. Being a proud people, they were offended by our attempt to be the "great white father". They further went on to state that in some fields, architecture, as an example, they were superior to us. They asked whether there was anything that I could do to correct the situation. I had never had any contact with the government and certainly not The State Department, however, I said that I would try.

Being a brash kid, nothing seemed impossible and it wasn't long before I found myself on a train to Washington. Since I was now making $35.00 a week from Intercollegiate Broadcasting Station Representatives, I had the funds to make such a trip. The entire expense was my contribution as I never had an expense account or salary from IBS. On arriving in Washington, I took a cab to the State Department. I presented my problem to the receptionist at the desk and during the next two hours I was sent from one office to another to discuss the situation of Inter-American broadcasts. I didn't seem to be making any progress and I soon gave up in disgust, although everyone treated me with respect. This process is known as the "revolving door treatment". I determined that there must be a better way to accomplish my purpose and I headed for the "hill". On arriving at the Senate Office Building, I inquired as to which committee was in charge of the budget of the State Department. I was referred to Senator Ralph Flanders of Vermont who listened to my story with great interest. My proposition was simply to invite the South Americans to participate in the State Department broadcasts that were continuously sent at them by inviting them to return broadcasts to us. These programs would then be rebroadcast over our college stations and any other standard broadcasting stations who wished to participate. In that way the Latin Americans would feel that they were a part of the broadcasting project and we would create goodwill instead of ill will. Senator Flanders referred me to Senator Fulbright who then sent me to see Senator Benton of Connecticut. Senator Benton had previously been connected with the State Department and his former secretary whom he referred to as "little brown eyes" was still there. After I outlined my plan in detail, he phoned her and asked to speak to one of the top administrators. In his conversation he stated, "I have a Mr. Bloch in my office who is connected with those college radio stations that are on many of the ivy league campuses. He had an idea which I think has great merit. Would you set up a full meeting tomorrow morning as Mr. Bloch must get right back to New York. The decision was made to set the meeting for 10:00 A.M.

The next day at 10:00 A.M. sharp I was again at the State Department but this time the red carpet had been rolled out and I was met at the door and escorted into one of the large meeting rooms. I must say that I was somewhat shaken to enter a room with over twenty top State Department people present. Again I presented my idea and was questioned by a number of the group. I was thanked and was told that my proposal was good and that some action probably would be taken. I later learned that within 36 hours after that meeting cables had been sent to fifty countries describing my idea. The request was made that the various ambassadors make inquiries in their countries regarding the possible interest in exchange radio programs. We did get some interest from several countries but the State Department never really followed through. I learned an important lesson, however, and that was how to obtain action from a branch of government. One will have a better chance of success in obtaining change in government, not by approaching the agency of government involved, but by taking up the matter with the appropriate committees of congress which control the budget for that agency.

International House, on a regular basis, invited prominent individuals to speak at Sunday afternoon meetings or at the Sunday night suppers. It was at one of the Sunday afternoon meetings that I first met Eleanor Roosevelt whom I learned to admire for her practical and humane approach to the problems of the day. She was a modest woman who never "put on the dog". I well remember her arrival on that Sunday afternoon. The Director and the staff of International House were all at the side entrance of International House awaiting her arrival. They were looking for a chauffeur driven car. She was a little late. As the group looked anxiously for her approach, suddenly a Fifth Avenue bus stopped and off stepped Eleanor Roosevelt and her friend Mary McLeod Bethune. On the two Sunday afternoon presentations at which she spoke she brought Mary Bethune who was an outstanding Black leader, a founder of Bethune-Cookman College. Mary Bethune referred to her own looks in uncomplimentary terms but, as far as I am concerned, she was a great fighter for social justice of which there was little in those days of segregation. In a later year, 1955, I had the great privilege of hearing her final speech at a Moral-Re-Armament meeting in Washington. The speech which she entitled "The Crowning Glory" was completely extemporaneous. The previous speaker, an old lady from Richmond apologized to Mary Bethune for her family who had been slave owners and with tears streaming from her eyes, she went over and kissed Mary McLeod Bethune. There was hardly a dry face in the audience. Mary Bethune opened her speech by saying that she was there against her doctor's orders but that she would not have missed that moment regardless of any consequence. She stated, "This is the crowning glory of my life and it is my hope that through acts such as you have just witnessed, that this will become a better world for us all." She died a few months later.

I met Eleanor Roosevelt on two other occasions, both of which involved a broadcast which was to be carried over the Muzak station, W47NY and picked up by our member college stations in the area. To make initial arrangements she invited a few of the IBS executives to her Washington Square apartment which was a large penthouse. The furniture was victorian and the living room was rather somber. The apartment reminded me of my aunt's which was very old fashioned and conservative. I always enjoyed being with Mrs. Roosevelt as she always had a good story to tell and she was a genuine person interested in helping people. Her subject for the broadcast was to be the contribution of college students and the American colleges in the war. The program was directed to undergraduates, Army and Navy trainees and other residents of the campuses.

The next day four of us picked up Mrs. Roosevelt by cab at her apartment and drove her to the studio at 70 Pine Street near Wall Street where the broadcast was to be transcribed. One of our group sat in the front seat of the taxi and three of us and Mrs. Roosevelt sat in the back seat. Mrs. Roosevelt was a tall woman and it was not easy to fit four of us in that taxi's rear seat. On the way to the radio station she told us about the problems that she was having with Westbrook Pegler who was continually insulting her in his column. She told how she had learned to write her column, "My Day". It seems that she had an agreement with the newspapers which allowed them to cut her column if they did not have enough space. Many of them cut out parts of her column making her look silly. She devised a method to write the column in such a way as to make cutting it almost impossible. The paper was not allowed to add to the column in any way.

When we arrived at 70 Pine Street, a skyscraper, we entered the building and went to the elevators where the uniformed elevator attendant saluted her. On entering the radio station, Mrs. Roosevelt was ushered into the studio. We sat in an adjoining studio which was open to the control room in which the engineers were cracking "Eleanor and Franklin" jokes much to our embarassment. We knew that she never heard this low grade humor but we were shocked at the poor taste exhibited.

On November 30, 1942 we tied six FM stations together to cover the Mid-Atlantic and New England regions for a series of programs entitled "You, the War and the Future". The originating station was W47NY, New York and was relayed to W2XMN, Alpine, N.J., W65H, Hartford, W43B, Boston, W39B, Mt. Washington, New Hampshire and W57B, Schnectady. The college transmitters picked up the broadcasts from one of the six FM stations by FM receivers which were installed at each of the college stations. Campus stations participating included those located at Brown, Columbia, Connecticut, Hamilton, Harvard, Rhode Island State, Princeton, Dartmouth, Union, Wesleyan, Williams and Yale. Programs were re-broadcast by long wave to the campus audiences.

The series "You, the War and the Future" consisted of a series of lectures given by several authorities in the field of current events. As reported in "Radio Daily" of November 30, 1942, "The first speaker was William T. McCleery, former executive editor of AP Feature Service and editor of "Picture News", Sunday feature section of the newspaper "PM". Under the direction of Leslie Katz, program director of the Intercollegiate Broadcasting System, each broadcast analyzed information from authoritative sources regarding the position of college students and young men of college age in the war effort and the post war era."

In 1942, at International House, I met a very interesting fellow, Edward Bindrim, who was a mechanic of sorts. He sold me on the idea of making an investment in a 1932 Ford which he had seen for sale in a junkyard in Brooklyn. The price was $35.00, quite steep for me, but I pooled my resources and made the purchase. In those days of gasoline rationing we were given small monthly gasoline allotments but since I used the car for business I was given a special allocation. Now the proud owner of a car, I decided to make a tour of the IBS member stations in New England and Joe Weed agreed to give me an expense account for the purpose.

Leslie Katz, IBS Program Director, and I set out on our trip and almost immediately we had tire trouble, blowouts constantly. To make matters worse, new 1932 style tires were not available and, although we had one spare, all we could do was have repairs made along the way. He and I arrived at New Haven, visited the Yale station and had a most enjoyable time. After spending the night at one of the dorms, we started out the next morning for Providence and promptly had another blowout. After much time wasted, we arrived at Brown, took a tour of the Brown Network and proceeded to Cambridge where we spent the night at Harvard. The station managers had many questions about IBS and we were able to brief them on many new developments such as exchange programs. National advertising revenue was flowing into the treasuries of the college stations and everyone was pleased with our efforts. The directors of the Harvard station were enthusiastic about our exchange programs between IBS colleges which Leslie Katz had organized and directed. The Crimson Network had become a major extracurricular activity at Harvard, as had most of the other stations on their campuses.

After spending the night in a Harvard dorm, we started out for Williamstown, Mass. and Station WMS, Williams College. Again we had a blowout, put on the spare and kept chugging along. As we approached the top of a small mountain, the car began to bounce and jerk. Upon investigation we found that a rear tire had split and the red tube was blowing out of the crack in the tire. Our spare had previously blown and we had no time to have it repaired. We had visions of spending the night on the top of the mountain in a disabled car. I was driving and told Leslie that we had no other choice but to try to make it to a gas station. I released the brake and at about ten miles per hour coasted down the mountain. Seven miles later we arrived at the bottom of the mountain, and as luck would have it we coasted into a garage which was just about to close. Nevertheless our tires were repaired and we hobbled into Williamstown.

Williams College is situated in a beautiful mountain setting and seemed isolated from the world after the heavy traffic and congestion of New Haven, Providence and Boston. We spent an enjoyable and restful evening after a fine dinner at The Inn. The Williams station had been one of IBS's most loyal supporters through good times and hard times and we were most appreciative. George Goldberg, Williams graduate, had helped me land our first national advertising account, The Standard Oil Company of New Jersey, ESSO. Because of the isolation of the college and the lack of competition from other broadcasting stations, the Williams College station had become the principle source of entertainment on campus for both those who received and those who produced the station's programs.

Our trip completed, Leslie and I returned the next morning to New York without further blowouts. The total number of blowouts during the trip was seven. My investment of $35.00 for the car had paid off as the trip was a great success.

Lest one think that a salary of $35.00 per week would prohibit one from enjoying life's little extras, I recount the following:

On arrival at International House, I parked my car on the street. Even in those days parking spaces were hard to find but I usually found a spot within a block of International House. Edward Bindrim was there when I arrived and he described his latest exciting project. In the same junk yard where he had found my car, he had purchased a truck body, windows and other parts necessary to build a house trailer. It was six months later when, at his invitation, I visited the junk yard to see the house trailer which he had constructed. During the period of construction, junk cars had been piled

65

in front of it and it took another month before he could get it out of the yard. In the meantime he went to the license bureau where they questioned him on the make and model of the house trailer. His answer: A Bindrim-1943. The next weekend I went with him again to the junk yard and he hitched the trailer to his car and we drove to the Ramapo Mountains in New Jersey, a distance of about fifty miles. We arrived at a campground which was bounded by a beautiful mountain stream and after paying our camping fee for a month, we parked, set up our camp stove and prepared to enjoy the weekend. The trailer consisted of a kitchen sink, toilet, shower with plenty of shelf and storage space but very little floor space. We had army cots for sleeping but found that two cots completely took up every inch of floor space so when we set them up, we found that it was impossible to enter the trailer by its door. That didn't bother us, we entered by the window. We had little sleep that night as it was hot and cramped and the next morning after a swim, we headed back to New York leaving the trailer.

To make a long story short, I ended up buying the trailer for $50.00; why I will never know. The next month I asked Edward if he would drive with me to the campground in my car and help me hitch the house trailer to the car. When we arrived at the campground we hitched up the trailer and proceeded to drive out but the owner, who, we found out later, was the new owner, pursued us ordering us to halt or he would call the police. We towed the trailer across an open field to the road. A few miles down the road the police stopped us but since we proved ownership of the trailer, there was no charge against us and our fee at the trailer park had been paid. It seems that the old owner included my trailer in the sale of his park. The strain of pulling the trailer was too much, however, for my car and just as we were in the middle of a long bridge approaching New York, the water in the radiator started boiling and the car stopped. It was a Sunday afternoon with about a mile of cars behind me on a three lane bridge. I took off the radiator cap and the water spurted three feet in the air. For such occasions I carried a box of baking soda which I had been told would help, so in the midst of honking and utter confusion I waited for the car to cool and then poured baking soda into the radiator. That was the only time that I was ever responsible for such a traffic jam.

Limping back to New York, I parked the car and the trailer in the parking lot at 125th Street down the hill from International House. I made a deal with the parking lot owner to sell him the trailer for $75.00 and the Bindrim-1943 ended its life as the shelter for the parking lot owner who installed in it a stove and made it his second home. Waste not, want not.

On weekends during the Spring, Summer and Fall a group of us from International House would take a bus and head for the Ramapo Mountains or Bear Mountain State Park up the Hudson for a weekend of hiking, camping and swimming. Our group consisted of from ten to twenty students of many nationalities. On returning we all went to the Chinese restaurant on Broadway for a hearty meal. We rented a cabin at the Bear Mountain State Park. The bus would let us off at the park entrance and we would hike the two miles up the mountain to our cabin carrying our provisions. Because of gas rationing, cars were scarce and we virtually had the park to ourselves. It was during those camping weekends that I really developed a great interest in foreign students. In later years I became a publisher of educational material for schools which would give a true picture of foreign people and their cultures. Much of the material which I later developed was written by natives of the featured countries.

Back row (l-r), George Abraham, David Borst
Front row (l-r), Naomi Fine, Louis M. Bloch, Jr.

Back row (l-r), Naomi Fine, New York Office, Arthur Ward, Hamilton
Front row (l-r), John Metzger, Ohio University, Bruce Bernstein, Wesleyan, Walter Wager, Columbia

Back row (l-r), Abraham Siegelman, R. I. State, Gilbert Cullen, U. of Maryland.
Front row (l-r), Raymond Page, R. I. State, Robert Harper, Haverford, George Reynolds, U. of Maryland.

Back row (l-r), Art Newman, Cornell, Sheldon Sprague, Swarthmore
Front row (l-r), Bob Kieve, Harvard, Lyn Granot, Swarthmore, Peter Binzen, Yale

CHAPTER VIII

COLLEGE RADIO COMES OF AGE

College radio stations formerly the laughing stock of the radio industry were slowly being accepted as an important media for national advertisers. I continued to attend the Friday luncheons of The Radio Executives Club where I met important people in the radio and television industry. Also attending the luncheons were account executives from advertising agencies, many of whom I knew as they were time buyers of the IBS stations. Included among my customers were such prominent agencies as Batten, Barton, Durstine and Osborn, J. Walter Thompson, William Esty & Co., Marschalk & Pratt, Mac Farland-Aveyard and Maxon, Inc.

At one of these luncheons I met Horace Titus, the son of Helena Rubenstein. His advertising agency, Advertising House, represented her account. He invited me to visit his office and showed me his new project which he called "The Color Spectroscope". His plan was to decorate apartments and homes to match the woman's complexion. He had made arrangements with manufacturers of rugs, drapes, furniture, curtains and pictures to develop special color combinations for different complexions. I sold him a spot campaign which was to run on all of the IBS women's college stations. He took me over to the Helena Rubenstein executive offices on Fifth Avenue and introduced me to his mother who was in her eighties at the time. At Christmas he invited me to a cocktail party at which were present people from radio, the press and other advertising agencies.

More advertising accounts signed including Finchley and Brooks Brothers Clothes, The Biltmore Hotel and Beechnut. Then Readers Digest ran a test spot announcement campaign over the IBS stations at Princeton, Wesleyan, University of Connecticut and the University of Alabama.

I conducted a survey for the R. J. Reynolds Tobacco Company who sponsored "The Camel Campus Caravan", three weekly swing music programs over the IBS stations. It was discovered that on our campuses, 98.6% of the students had access to radios. The college station was the favorite over all other stations in 64.5% of those interviewed.

George Goldberg of station WMS, Williams College, the son of Rube Goldberg, sometimes helped me solicit accounts. I will never forget the day when George and I went to the offices of a large record company. We asked for the manager and the receptionist disappeared. She returned to tell us that the manager was out of town. He did not realize that we could see him seated at his desk as a mirror reflected from his room. That night George asked his father, Rube, to draw a cartoon of the incident and the next day we returned to the office and asked the receptionist to deliver Rube's cartoon to the manager. The cartoon showed an harassed manager with his hair flying in all directions telling his receptionist "Tell Them That I Am Out of Town", signed Rube Goldberg. The manager apologized and we all had a good laugh, but we never did get the account.

I had much better luck with RCA Victor whose headquarters were in Philadelphia. After arriving at RCA, I was ushered into the sales manager's office. He explained that it was his job to promote RCA Victor records. I proposed that RCA sponsor both popular and classical RCA records over the IBS stations. To do that, however, we would require that RCA donate collections of both popular and classical RCA records to each college station. The deal was consummated and RCA purchased three weekly programs on all available IBS stations. They donated to each station the Standard RCA Broadcast Album, comprising approximately 800 classical recordings and 300 popular pressings. Both groups were to be supplemented each month by shipments of the latest RCA releases. Two nights a week RCA sponsored a program of popular favorites entitled "The Campus Tune Parade" and once each week a forty-five minute classical music program. The contract with RCA Victor was no doubt, the most important that I signed for the IBS stations. Not only did the stations acquire a lucrative contract, but also a great popular and classical record collection.

Bob Girvin in 1945, as a freshman, was the director of classical music programs for The Harvard Crimson Network. He recently told me that the RCA Victor record collection filled an 8' x 10' room from the floor to the ceiling with RCA Victor albums with new records arriving monthly. Today Bob Girvin is a bank executive in Cleveland.

I signed a five minute news program with Old Gold, a spot announcement series with The New York Herald Tribune and, for the Cornell station only, a spot announcement series with The New York Telephone Company.

In the Fall of 1943 I moved out of International House to Larchmont in Westchester County where I found a room in a private home. Living in Larchmont was an interesting experience and I enjoyed driving to Long Island Sound to watch the sunset and the yachts and sailboats. Since I had a car, such as it was, I was able to drive to the station each morning and take the train into Grand Central Station which was only a block from the IBS office at Fifth Avenue and 42nd Street. Driving a broken down 1932 Ford in the affluent community of Larchmont was an experience. One day I was stopped by the police who told me that a part of my muffler had dropped off and another time when they told me that a section of my fender had become dislodged. When I drove to the railway station to commute into New York I found myself surrounded by Cadillacs and other fine cars.

One Sunday noon I was invited for lunch to my uncle's country club near Rye. On arrival I drove to the parking lot past the terrace where the wealthy crowd had gathered. My uncle told me later that everyone thought I was the gardener rattling into the drive.

After living in Larchmont about a year, I moved to an apartment with a friend near International House. We had a front room and a bedroom but shared a bath with a Russian emigrant who lived in the back suite.

During the Summer of 1944 two friends and I drove out to Long Island and found a cottage for rent in Bayville on the beach at the end of the road. Bayville is located on a spit of land across from Oyster Bay and the Theodore Roosevelt home. We rented the house for the Summer and commuted to New York each day. As we were at the end of the bus line, the driver would stop and have coffee with us and then drive us to the station about five miles distant where we would board the steam train for New York. After a change to the electric train at Jamaica, we proceeded to the Pennsylvania Station. The trip took an hour and a half each way but it was well worth it, as when we arrived home in evening, we were able to change into our trunks and take a fast swim in the Sound by just running across our front yard which was the beach.

Living in the cottage on Long Island Sound was most enjoyable and on the week-ends we had visitors in droves who took advantage of our cottage to get away from the heat of New York.

In November of 1944, The General Electric Company bought five minutes, five times a week on thirteen IBS stations to sponsor the "GE Campus News". This most popular program was heard daily at Harvard, Yale, Cornell, Alabama, Columbia, Brown, Williams, Wesleyan, Haverford, Bryn Mawr, Union, Stephens and Radcliffe.

After the program had run for four months, I conducted a survey for General Electric to determine the results of their advertising. The result of the survey appears in the appendix.

We had two advertisers sponsoring time signals, Ingersoll and The Gruen Watch Company. Gruen bought time on all IBS stations and featured their sponsorship of time announcements over the IBS stations in their publication "Gruen Time" which went to all Gruen dealers. In "Gruen Time" they published a feature as follows:

HISTORIC COLLEGE HALLS
Resound to Gruen's Time Jingle

"The venerable ivy-covered walls of many of our most honored and historic institutions of learning are 'hep' to this generation of 'solid', sadddle-shod, students whose daily schedules are timed by the merry 'tick-tock' of Gruen's famous Time Jingle.

Sixteen colleges and universities with a total enrollment of 35,000 students are receiving the Gruen Time Signal over The Intercollegiate Broadcasting System. The daily routine of these ambitious young men and women is being regulated today by Gruen Watch Time.

College class rooms are no longer idle during long vacation months, nor do they hum with the erudite recitations of bespectacled postgraduates boning for another degree. College as of 1945 is a round-the-calendar program of intensified concentration by the pick of the nation's youth, many of them Army and Navy trainees, telescoping unheard of courses into a precision-timed schedule of hours, weeks and months. They are the business men and women of the future, the parents and the homemakers, citizens of a world to be dedicated to peace, and they go to their classes, to meals, and to study by the time that's famous for precision. In twenty-two seconds oft repeated, the name of Gruen is being tunefully woven into the warp and woof of their busy, pleasant days-days that will glow for years to come with unforgettable college memories.

The colleges and universities receiving the Gruen Time Jingle are distinguished for the many and impressive achievements of the great men and women whose names are found among their alumni, and for their own progress in the field of education. They are Alabama, Brown, Bryn Mawr, Bucknell, Columbia, Cornell, Harvard, Haverford, Princeton, Radcliffe, Stephens, Union, Wellesley, Wesleyan, Williams and Yale.

The thousands of students who regulate their daily activities by Gruen Watch Time are an important addition to the millions who hear Gruen's catchy Time Jingle daily from coast to coast, and to the many thousands of listeners of local radio programs sponsored by retail jewelers who use this prestige building time signal."

In 1945 the Federal Communications Commission announced that hearings would be held in Washington regarding the allocation of FM channels.

George Abraham contacted John W. Studebaker, United States Commissioner of Education regarding the hearings and Studebaker requested George's help in drafting the proposal for educational FM. Up to that point education had received very little consideration from the FCC as almost all channels had been given to the commercial interests.

Even though there was little interest in FM stations at the time, it was realized that it was important that education be allocated channels for future use. The request was made for 20 FM channels and 2 relay channels to be devoted to education.

Commissioner Studebaker contacted the currently operating educational broadcasters requesting that they testify at the FCC hearings. George asked the IBS colleges to send their most distinguished alumni. Since IBS included the Ivy League colleges there were a number of governors, members of congress and business leaders who could be contacted. As a result of this dual effort 33 distinguished individuals testified for education at the FCC hearings. As a result, all 20 requested FM channels were given to education plus the two relay channels. An FCC member later stated that there had never before been such a lobby for education before the FCC.

Although invited, I did not attend the hearings. I soon, however, heard the reaction of the FCC allocation of the FM channel in a phone call made to my New York office from an American Telephone & Telegraph Co., vice president. He invited me to discuss the matter at the New York AT&T office the following day. When I arrived there I was introduced to six AT&T officials who tried to convince me to relinquish the two relay channels which had been given to educational radio. I did not give AT&T much satisfaction.

It seems that two AT&T vice presidents, on the train to Washington, had been reading the release from the FCC regarding the new education frequency allocations and were shocked by the fact that AT&T did not get the two relay channels which they had expected. When they arrived in Washington, they called George who gave them no satisfaction but referred them to me at the IBS New York office.

The FCC reserved 20 channels-between 88 and 99 MHz for noncommercial use. The reserving of 20 channels for eductional use was a precedent-setting move as it established the principle of withholdng part of the broadcast frequencies from commercial use. It was a great victory for educational radio for which John Studebaker had been fighting for so many years.

As of July 31, 1980, 1,063 licensed educational FM stations are on the air, several of which operate at the power of 50,000 watts.

In February, 1946, I resigned as Business Manager of the Intercollegiate Broadcasting System and Manager of Intercollegiate Broadcasting Station Representatives and returned to Cleveland. Unfortunately the New York office lasted only one year after my leaving.

CHAPTER IX

THE PIONEER "GAS PIPE" COLLEGE RADIO STATIONS

Those "gaspipe" campus radio stations which commenced broadcasting prior to 1946 are considered pioneer college radio stations.

Included in this section are brief histories of the pioneer stations except those at Brown, Columbia, Wesleyan and Williams which are previously described. The few stations not described did not submit the necessary informaton by press time and they will appear in the second edition.

George Abraham and David Borst, the founders of college radio, in the control room of The Brown Network.

UNIVERSITY OF ALABAMA,
(Bama Radio Network, BRN)

The college radio station, Bama Radio Network (BRN) at the University of Alabama commenced operation the first week of January, 1942. Ten low powered transmitters, located at strategic points around the campus, each with a 250 foot radius, broadcast local and national news programs, dramatic programs, and interviews with visiting celebrities. Service included dormitories, fraternities and sororities. Personnel of BRN included Roy Flynn, President; Robert Whitehurst, Technical Manager; Lem Coley, Director of Public Relations; Beth Barnes, Program Director and Frank Eddens, Office Manager. Faculty advisors were Professor John S. Carlile, Director of Radio Education, and Professor A. A. Kunze, Instructor in Electrical Engineering.

Because of the war and the shortage of men, many of BRN broadcasts featured women. A typical daily program schedule follows. BRN broadcast only one hour per day:

5:00 p.m.	Rhythm Girls
5:15 p.m.	Jill and Jean
5:30 p.m.	Eva Symonds
5:45 p.m.	News
5:50 p.m.	Previews

The Control Room at BRN

79

ANTIOCH COLLEGE, YELLOW SPRINGS, OHIO
(Station ABS)

The idea of a campus radio station at Antioch was conceived in the Fall of 1939 when Bruce McPhaden, then Community Manager, began playing records in his room in North Hall on a wireless turntable he had bought. This turntable, without use of wires, broadcast to radios within 260 feet of his room. Soon other boys in the dorm began listening to his broadcasts and ABS was born.

Several months later Community Council appropriated money to buy equipment for a more complete station and a studio was set up in the Music Library. Music, obviously, was still the station's mainstay, with an occasional quiz or feature program thrown in.

In 1940 the equipment was rebuilt and moved to a new, permanent, soundproof studio on the fourth floor of the Science Building.

In 1941 the setup remained essentially the same, except that there was better organization. A definite schedule was adhered to every night between 7:00 p.m. and 10:00 p.m. This included campus and foreign news, sports reviews, skits and both popular and classical music. A new feature was a regular broadcast of Community Council meetings. In charge in 1941 were Hardy Trolander and Dave Case, who rebuilt the equipment.

A typical program schedule was as follows:

> 7:00 p.m. Campus News
> 7:15 p.m. Sports Review
> 8:00 p.m. Music
> 8:30 p.m. Swing Music
> 10:00 p.m. Sign off

BUCKNELL UNIVERSITY, LEWISBURG, PENNSYLVANIA (Station WBRW)

In December of 1942, Bill Roos '43, head of The Bucknell Radio Workshop, completed work on a small transmitter which was to broadcast recordings and campus news to the University dormitories. A trial broadcast was completed. Roos was drafted and it was not until March 24, 1944, through the help of Professor McRae, that the station commenced operation.

In November of 1945, station WBRW moved to a new location, 420 South Seventh Street, a small two story frame building directly behind the Sigma Chi house. The station consisted of a studio, control room, announcing booth and vestibule on the first floor and an office and workshop on the second floor. Profits from "commercials" financed the construction of a new transmitter by the engineering students under the direction of Professor James Richards of the physics department and Robert Kelly, a chief engineer of WBRW. Station WBRW broadcast 16 hours weekly and could be heard at 640 on the radio dial.

The Radio Workshop not only operated WBRW but also broadcast plays, talent shows, quiz programs and variety shows over their local radio station WKOK.

In the 1945 issue of L'Agenda, the Bucknell yearbook, The Radio Workshop was described as follows "The Radio Workshop is one of the youngest, but most rapidly growning organizations on our campus. Students who are interested in radio have the opportunity of learning a great deal about the technique and operation of radio, since the studio is organized and managed entirely by the students themselves."

UNIVERSITY OF CONNECTICUT, STORRS, CONN.
(Station UCBS)

The Husky Network, as the University of Connecticut station at Storrs, Connecticut was called, commenced operation April 8, 1940. Broadcasts were scheduled on Monday, Wednesday and Friday from 7:15 p.m. to 8:30 p.m. The studio was set up in the Community House through the cooperation of Miss Etta M. Bailey, house director. Regular features of the Husky Network were a campus news review by Tom Leonard, latest sports news by Herb Peterson, weekly performances by the Radio Players and request programs handled by Tony Sarrett and Edward Scott-Smith. Edward Temkin was in charge of a classical program called "Musical Gems". Program directors of the Husky Network were Stanley Markowski and Valery Webb. The President of the Husky Network was La Vergne Williams and engineers were Tauno Ketonen and Ned Hines.

Until the Fall of 1941 the station relied entirely upon funds granted by the Student Senate. In September, 1941, when they secured an advertising contract with the R. J. Reynolds Tobacco Company which sponsored The Camel Campus Caravan through IBS, UCBS became a self-supporting station. Their studio was moved to old fraternity rooms of Tau Epsilon Phi in Hall Dormitory. In February, 1942 the name of The Husky Network was changed to UCBS, The University of Connecticut Broadcasting System, and the hours of broadcasting were increased to twenty per week. The studio in Hall Dormitory was improved and equipped with a new microphone and new monitoring equipment was installed. UCBS covered ten major dormitories and fraternity houses on Connecticut's campus.

A typical daily program over Station UCBS was as follows:

5:00 p.m. Swingland
6:00 p.m. Danceland
6:30 p.m. Slip Minous Varieties
6:45 p.m. March Time
7:00 p.m. Fred Waring
7:15 p.m. Freshman Sports
7:30 p.m. "X" Fraternity Hour
8:00 p.m. Musical Jems
8:30 p.m. Campus Hit Parade

CORNELL UNIVERSITY, ITHACA, NEW YORK

(Station CRG)

In 1939 The Cornell Radio Guild proposed construction of a wired radio station run by and serving the students exclusively. The Guild was offered a room in Willard Straight Hall for the studio and office. There was a condition on the use of the room and that was that it must be soundproof with a maximum of neatness and decor. Not a nail might touch the "Straight's" walls and no paint might even suffer a scratch. The hardy engineers of that first station were forced to build a room within a room. This room was divided into two parts, a studio and a control room with a large rectangular glass panel between the two rooms. Small transmitters were located in dormitories, fraternity and sorority houses.

Station CRG presented its first program on November 1, 1940, and on December 12, 1940, began its commercial schedule of four hours of programming daily. Station CRG had a membership of 200 students. Programs consisted of drama, news interviews, sports and recorded music.

In October, 1946, CRG became WVBR, the Voice of the Big Red. In 1958 WVBR made the jump from a wired AM station to a licensed FM station. Today WVBR-FM operates at a power of 3,000 watts and features "album oriented" rock music.

DARTMOUTH COLLEGE, HANOVER, N.H.
(Station DBS)

The Dartmouth College Broadcasting System, station DBS, commenced broadcasting the week of October 29, 1941. The staton was under the directon of Station Manager William Mitchel from Teaneck, New Jersey and his eight man directorate. The station covered all of the college dormitories, the faculty Graduate Club and the home of President Ernest Martin Hopkins. Small transmitters were placed in the various buildings covered and these were connected to the main control room by leased telephone wires. Because of the fact that Hanover was long the victim of much static and interference making other radio reception difficult, DBS was welcomed because of its clear reception.

Originally DBS was financed by the college, however, advertising was soon obtained and this made it possible to pay back the original investment, and to meet operating costs and maintenance expenses.

The station began with a three hour per day broadcast schedule which included everything from jive to weather reports. Some DBS time each evening was given over to Dartmouth professors and one of the most popular programs was entitled "Meet The Professor". One of the most spectacular accidents occurred when Professor Bear of the Psychology Department was presenting his program on "How to Study". A voice from the control room both startled and pleased his radio audience with, "Dammit, I've heard all this somewhere before", and his assistant's retort, "Sure, dope, you heard it this afternoon at rehearsal." The youthful commentators and announcers learned to rise to such occasions gracefully.

GEORGETOWN UNIVERSITY, WASHINGTON, D.C.
(Station GBS)

Station GBS made its debut on November 12, 1941. The initial program was heard by 800 students in dormitory rooms and an invited group in the auditorium of the Speech Institute.

Carl Bunje, president of the students' radio club announced that the new Hilltop station was "dedicated to a greater Georgetown". University officials took part in the half-hour program and emphasized that the unnamed alumnus, whose contribution made the station possible, had in mind the preservation of "the dignity of the English language."

The Very Rev. Arthur A. O'Leary, S.J., president of the university, sent a congratulatory message in which he predicted the new campus activity will "develop into a splendid link between the faculty and the student body and will be a practical factor in the daily life of the university."

The dean of the College of Arts and Sciences, the Rev. John E. Grattan, S.J., who was helpful to the students in promoting the station and the Rev. Edmund A. Walsh, S.J., Regent of the School of Foreign Service, who took an active interest in it from the start, participated in the broadcast.

Father Grattan reminded that Pope Pius XII, in praising the new Vatican radio station a year ago, referred to it as "This marvelous bridge". "I hope that not a favored few, but as many as possible of the students will make use of this bridge," Father Grattan said, "and find it a means of establishing closer contacts with their fellow students."

Coming events at Georgetown were broadcast every morning Monday through Friday during the school year to keep the student body informed of extra-curricular activities. On Sundays from 10:30 a.m. to noon there were special broadcasts of popular and classical music. Evening programs commenced at 8:30 p.m. and included drama, swing music, round table discussions, sports broadcasts and campus news.

Station GBS had a short wave listening post and when, on Sunday December 7, 1941, the first bombings of the far eastern war began, technicians and announcers monitored a battery of short and long wave receivers in the darkened studios of GBS, recording important comments and speeches from the ether, and then rebroadcast them to the student body in the morning.

GBS broadcast commentaries of the war news were by George Cain, editor of THE HOYA. The description of the terrain and military objectives in Manila were by Carl Bunje, president of GBS whose home was The Philippines.

On February 23, 1942, when GBS commenced operation after mid-term, the daily program schedule was increased. A typical day's broadcast signed station GBS on the air at 8:15 a.m. with a program of swing music designed to open the eyes of the early riser. From 8:30 a.m. to 8:45 a.m. the morning campus news program went on the air, followed by a station sign-off. At 5:00 p.m. in the afternoon GBS broadcast its second program, a full hour of classical music. From 8:30 p.m. to 9:00 p.m. each weekday evening, Monday through Friday, the feature show of the day was broadcast. On Tuesday and Friday, this show was sponsored by Beech-Nut. On Wednesday the feature was "Spotlight", the special war coverage program. "Blumack Varieties", a dramatic program was featured at least once a week.

HAMILTON COLLEGE, CLINTON, N.Y.
(Station WHC)

The success of Station WHC, Hamilton College was mainly due to the efforts of one man, Wentworth W. Fling of the Hamilton College faculty. He arrived at Hamilton in 1938 to teach the romance languages and stayed to enlarge his scope of activity so that, in addition to teaching French and Spanish classes, he was director of phonetics and organized and designed the campus radio station.

Station WHC, located in the basement of Root Hall, commenced broadcasting in the Fall of 1941. It had two studios, a control room, a business office, a workshop and a transmitter room. Fling designed it all and helped to build it. He was quoted as saying "We do everything at WHC that is done in a commercial station. The students write scripts, sell advertising, plan the daily programs, announce and take part in plays." Fling felt that when students finished their association with the station they would be prepared to take jobs in the radio field.

Even though Station WHC served the smallest college, this pioneer station boasted the largest percentage of the college enrollment on its staff.

Frank Miller '45, Hamilton College, chief engineer of the school's radio station WHC, is shown in charge of the station's control room.

HARVARD UNIVERSITY, CAMBRIDGE, MASS.

(The Crimson Network)

The Harvard Crimson Network commenced broadcasting on April 15, 1940. The station owned and operated by the Harvard Crimson, the college newspaper, covered all parts of the Harvard campus within reach of university heating pipes.

The inaugural program at 7:00 p.m. on Monday April 15, 1940 featured a talk by Dean Chase, the chairman of the Harvard Radio Committee and Loring Andrews, Program Director of the World Wide Broadcasting Foundation. The broadcast frequency was chosen as 800 on the radio dial. After the inauguration of The Crimson Network, the first program was "Jazz Lab", a program treating the development of types of jazz from its beginnings to modern swing and then followed by a classical music hour. Rounding up the first day's program was a summary of college news.

Instructions were issued to the Harvard students as follows: "Students will not have to make special attachments to the heating pipes to receive the college station with ordinary sets, however a ground on the pipes may hinder reception, whereas an aerial attachment to the radiators might improve the reception considerably".

The first officers of the Harvard Crimson Network were William W. Tyng '41, Executive Editor of the Harvard Crimson, Chairman; Charles W. Oliphant '41, Technical Director; Charles W. Davis '41, Program Engineer; Lawrence Lader, Program Director.

Reception throughout the University was far from perfect as many would be listeners, including President Conant, reported that they were unable to tune in the programs. Poorest reception was in Eliot and Leverett Houses and in general on the top floor rooms of the college buildings.

Students were informed by the station engineers that transmission difficulties would soon be worked out.

It was not long until The Crimson Network became an important extra curricular activity at Harvard. Broadcasts started at 8:00 p.m. and continued until 11:00 p.m.

For its first two years The Harvard Crimson Network operated without advertising due to the fact that the administration opposed it. It became evident that unless The Crimson Network was permitted to obtain advertising that it would be unable to continue. On February 2, 1942, after almost two years of operation without commercials, The Crimson Network went on the air with its first commercial, a spot announcement by the Beech-Nut chewing gum concern. From the revenue obtained from the Beech-Nut contract The Crimson Network could dissolve its debts, pay its small maintenance expenses and extend its broadcasting facilities to the Harvard Yard. The Network announced that it would limit the amount of time given to commercial announcements. "The Network intends very seriously to allow no advertising contract to alter its chief purpose —that of supplying its listeners with good, and with listenable programs", the executive board announced.

The Harvard boys did, however, run a number of "offbeat" programs. The first play in a series on the college and its history was entitled "Lo, the Butter Stinketh". The students, in the most part, were conservative about what they broadcast, but the Crimson Network did cause Dean Chase anxiety when it broadcast an interview with Ann Corio, burlesque star, on the art of the strip tease.

A typical program schedule of The Crimson Network was as follows:

9:30 p.m.   News Analysis by Robert Kramer '42
9:40 p.m.   Jack Lampl '42, swing piano

9:50 p.m. The Witness Stand
10:00 p.m. Concert Hall: Manuela deFalla, Nights
in the Gardens of Spain. Three Cor-
nered Hat Ballet.
10:50 p.m. Network Sports Review
10:55 p.m. College News'
11:00 p.m. Code practice

Today Harvard operates a commercial FM station
WHRB-FM with a power of 3,000 watts.

Charles W. Davis '41, Program Engineer for the Crimson Network, carries out
preliminary tests seated at his microphone.

HAVERFORD COLLEGE, HAVERFORD, PA.
(Station WHAV)
BRYN MAWR COLLEGE, BRYN MAWR, PA.
(Station WBMC)

The Haverford college station had its small beginning in the spring of 1941 when Henry H. Gray built a small transmitter in Founders Hall to broadcast to those in that dormitory. Programs were informal and at infrequent intervals and included recorded music and one radio play. Before the end of 1941, permission was secured to replace the old station with a new one with greater coverage. The new station was built during the summer of 1942.

On October 20, 1942, the Haverford College Radio Station, WHAV commenced broadcasting from studios in the Union. President Felix Morley, Professor Frederick Palmer, Jr. and John W. Clark, President of the Radio Club took part in the opening program. Programs were originally heard between 7:30 p.m. and 9:30 p.m., Monday through Thursday.

From the beginning Bryn Mawr students expressed interest and participated in the WHAV programs as actors, singers and technicians and worked behind the scenes as script writers.

On February 9, 1944 station WBMC at Bryn Mawr joined WHAV from their own studio located in Penn East. After the official christening with a figurative champagne bath given by Miss McBride, WBMC joined WHAV to operate together as a network. WBMC was built by the Haverford boys during the Fall of 1943.

The program schedule of the Haverford and Bryn Mawr stations was changed to the hours of 8:30 p.m. to 10:30 p.m.

KNOX COLLEGE, GALESBURG, ILLINOIS
(Station WKC)

Station WKC, The Siwash Broadcasting System, made its initial broadcast the first Monday in February, 1942. The station covered the 600 student population in the college dormitories and fraternities.

"The Knox Student" printed the following account of the first broadcast. "Fellows gathered in the recreation room of Whiting Hall Monday afternoon to join the girls in listening to the first program to come from the microphones of the theater studio of The Siwash Broadcasting System.

President Davidson represented the faculty on the first broadcast. Bob McClelland, president of The Siwash Broadcasting System and Bob Boyes, program manager, made short speeches appropos of the station's opening.

Special organ recordings of Knox songs in the Dr. James Mac Weddell style added a musical touch to the program."

Studio apparatus and dormitory transmitters were built and installed under the direction of Ted Pihl, a physics major at Knox.

Program schedules were 5:00 p.m. to 6:00 p.m. Monday through Friday and 7:00 p.m. to 10:00 p.m. Monday through Thursday.

An innovaton in the Siwash Broadcasting System's setup was its cooperation with Galesburg's radio station WGIL. WGIL carried a weekly sports summary from WKC, and Natalie Woodward's fifteen minutes of song, in addition to the Knox Playhouse program. WGIL and WKC combined to cover all home football games.

Other programs featured over WKC were the daily news summary and the sports roundup by Russ Freeburg.

Today WKC operates as WVKC-FM, a 10 watt educational FM station.

OHIO UNIVERSITY, ATHENS, OHIO
(Station WOUB)

Station WOUB, a wired-wireless station commenced broadcasting in 1942 under the supervision of the School of Dramatic Arts, with the urging and staffing of students. From makeshift studios in the theater balcony of the now-demolished Ewing Hall on the Ohio University campus, the station sent its signal over wires strung through the heating ducts leading to other campus buildings. The station operated on the frequency of 560 kilocycles.

Radio students complained of broadcasting interruptions because of rehearsals and productions in the theater and in April, 1947, operations were moved to a Quonset Hut in front of the University's Alden Library. It was here that the FM station was born. The Federal Communicatons Commission had just modified its regulations to permit low-wattage FM stations and allow greater latitude in days and hours of operation — changes that were attractive to student run stations. On the recommendation of a Radio Advisory Committee, the Board of Trustees approved an FM station in December, 1948 and on May 25, 1949 the FCC granted a construction permit. The station made its first official transmission on December 13, 1949, one of the earliest 10 watt student operated stations in the nation to do so.

Today WOUB-FM is transmitting at 50,000 watts as an educational station covering all of southeastern Ohio and northwestern West Virginia and its present location is in the multi-million dollar Radio Television Communications Building on College Street.

PRINCETON UNIVERSITY, PRINCETON, N. J.,
(Station WPRU)

Princeton's station WPRU, small but swanky, had its home in the dormitory room of H. Grant Theis '42, its creator and builder. The transmitter was in the basement of the same building. Next to Theis' room was a small storeroom used as a continuity-writing room. There was just room for four continuity writers to sit in a row. In addition to broadcasting from his room, Theis' station could originate programs at various points on the campus to which he had underground lines running through the steam tunnels. These remote locations included the gym and one classroom building. In the basement the station was connected with the university power lines which carried the programs throughout the campus. A member of the original staff, James L. Robinson '43 (an electrical engineer), helped guide the development of the station as chairman of its governing board for over twenty years after its founding in 1940.

In a poll taken by WPRU in 1942, it was found that Glenn Miller was Princeton's favorite orchestra with Tommy Dorsey second and Benny Goodman third. Following in order of preference were Jimmy Dorsey, Gene Krupa and Artie Shaw. Among the favorite WPRU programs were the news and sports broadcasts and "The Camel Campus Caravan". Every Friday night M. B Williamson '43 presented a full blooded radio serial entitled "Tiger Finkelstein in the 23rd Century."

A typical daily program schedule follows:

5:00 p.m. Jimmy Dorsey, Harry James, Artie Shaw
6:00 p.m. Sign Off
7:30 p.m. Bob Crosby
8:00 p.m. Classical Music
8:45 p.m. Vaughn Monroe
9:00 p.m. One Man's Opinion by Dick Duff

9:15 p.m. Alvino Rey
10:00 p.m. Classical Music
10:15 p.m. WPRU News. Five minutes of the latest
      campus news as gathered and prepared
      by the University Press Club
10:20 p.m. Charlie Barnet
10:30 p.m. Sign-off

In 1955, fifteen years after its founding WPRU obtained a license to build an FM transmitter and became the first student-owned and operated FM station in the country. It beamed a 250 watt signal within a radius of twenty miles, while continuing to broadcast simultaneously to the campus on the AM dial by carrier current. The call was changed to WPBR. In 1960 WPBR increased its power to 17,000 watts and became one of the most powerful FM stations in New Jersey. In 1962 WPBR became the first college station in the United States to engage in stereo broadcasting. In the mid-seventies the station had about one hundred undergraduates on its staff and had an audience of 45,000 in a five state area, serving its listening public with the same basic ingredients it had given its campus listeners in 1940 — music, sports, and advertising plus national news, public affairs programs, lectures, chapel services and live concerts in stereo.

H. Grant Theis, creator of WPRU, broadcasting from his dormitory room.

RADCLIFFE COLLEGE, CAMBRIDGE, MASS.
(Radio Radcliffe)

Radio Radcliffe commenced broadcasting July 6, 1943 with the following announcement by Ronnie Phoenix, the president of the newly chartered Radio Club. "This is Radio Radcliffe-Radcliffe's voice of the Intercollegiate Broadcasting System, 560 on your dial". She introduced Miss Sherman who officially inaugurated broadcasting activities with a brief talk. Joan McGowan also spoke a few words of welcome on the part of the students.

Then Ronnie Phoenix explained the genesis of the station from a gleam in the eyes of a few undergraduate visionaries to a studio on the top floor of the Sports Field House transmitting a signal over the Quad's electric light circuit five nights a week from 8 p.m. to 11 p.m. The remainder of the evening's program included a commercially sponsored broadcast with Harvard's Crimson Network, and a monologue written by Emily Jacobson.

Both the Harvard and Radcliffe stations featured nightly exchange broadcasts via a leased telephone wire. The Radcliffe News printed the daily program schedules for Radio Radcliffe. Radio Radcliffe's contribution for the exchange program was a half hour program of popular band recordings entitled "Swing Out". Two nights every week the Crimson Network sent to the Radcliffe dormitories a full hour program consisting of the assignment in Music I, one of the most popular survey courses offered at both colleges. Other programs included interviews, original plays and new programs.

Radio Radcliffe was built with spare parts left over from the early days of The Crimson Network thanks to the generosity of William Flook '44, manager of The Crimson Network.

When I visited Radio Radcliffe, the station was located in an area about the size of a large closet. The station was affectionately called RR^2 and the programs were enthusiastically received by the students.

A typical program schedule of Radio Radcliffe follows:

7:00 p.m. Swing Out (Broadcast to Harvard)
8:00 p.m. Concert Master (From Harvard)
9:00 p.m. Big Woman on Campus
9:15 p.m. Music From "Oklahoma"
9:45 p.m. Reading From Robert Benchley
10:00 p.m. Music To Read By

RHODE ISLAND STATE UNIVERSITY, KINGSTON, R.I. (The Rhode Island State Network)

The Rhode Island State College Network at Kingston, R.I. commenced broadcasting April 1, 1940. John Stasukevitch was the station's first President.

In a preliminary audience survey to determine program preferences, results were as follows: Number one by a large margin was swing music, with news, drama and sports following in that order. Top musical choices included Glenn Miller and Fred Waring.

Limited service of the station commenced April 1, 1940. Actual construction had begun in January, 1940 under the guidance of Professor Hall and was financed by a loan from the college and gifts from college organizations. The engineers devoted an average of fourteen hours a day for construction, sacrificing their vacations and spare time. Dave Borst from the Brown Network acted as their technical consultant.

Edwards Hall was the broadcast headquarters. The Rhode Island State College Network operated four hours each evening. Two men were always on duty, one announcer and one operator. Types of programs were music, drama, news and large scale broadcasts such as concerts.

Two technicians of the Rhode Island State College Radio Network. Left to right: John Stasukevich, president of the Network and Clifford Ely, chief operator.

SWARTHMORE COLLEGE, SWARTHMORE, PA.
(Station WSRN)

WSRN started in 1940 with $150.00 in cash and a box full of wires, tubes and panels and in 1946 had close to $5,000 worth of equipment. The studio occupied a space of 24' x 30' under the rafters of Old Trotter Hall. On your right, as you came in, was the central control room and behind it, the soundproof announcing booth. The major programs at WSRN were music from a large collection of popular recordings and a classical library of 4,000 records. Other programs included dramatic shows every two weeks, instrumentalists, news commentators (one conservative, one liberal), round table discussions and sports broadcasts. At Swarthmore about 90 students or 10% of the student body took part at one time or another in the WSRN broadcasts with about 30 active members serving on the programming, technical and business staffs. The station was a major extracurricular activity on campus and had many listeners.

Swarthmore, Haverford and Bryn Mawr formed a regional network and exchanged broadcasts two or three evenings a week. The programs were relayed between the colleges over leased telephone lines.

Actors receive cue from director as "Workshop" program gets under way in the studios of Swarthmore's Station WSRN.

Interrollegiate Broadcasting System

Founded at Brown University Feb. 18, 1940

Station Membership

These presents certify that

UNION COLLEGE RADIO CLUB

was granted class membership C

in the System _December 28 /941_

Under article III of the Constitution

George Abraham
CHAIRMAN EXECUTIVE COMMITTEE

Louis M Black Jr.
BUSINESS MANAGER

_____ FOR THE MEMBER STATION

_____ FOR THE MEMBER STATION

UNION COLLEGE, SCHNECTADY, N.Y.
(Station UBS, UCRS)

Station UBS, Union College commenced broadcasting from the Engineering Building on September 22, 1941 as a "wired wireless" station serving North College and the Kappa Alpha Fraternity. The station was constructed by the College Radio Club with the help of David Borst, Technical Manager of The IBS. Later the station covered the entire campus and its call letters were changed in the Fall of 1942 to UCRS (Union College Radio Society).

Featured on the opening program were President Fox, speaking for the college, Myron Mills, '42, president of the Radio Workshop, Robert Killian '42, president of the student body, Warren Perkins '42, president of the Radio Club, Armand Feigenbaum '42, editor of The Concordiensis and John R. Sheehan, program manager of General Electric stations WGEA, WGEO and W2XOY, who acted as master of ceremonies. Sheehan, in announcing the various speakers, told of experiences in the early days of radio, congratulated the organization on its achievements and offered fullest cooperation with the system.

UCRS broadcast a variety of programs including popular music such as Fred Waring, barber shop quartets and classical programs. Also all home basketball games were broadcast from the gymnasium. Fraternities participated in programs as well as did the V-12 Navy unit on campus.

This year, 1980, Union College is celebrating the 60th anniversary of W2XQ, an experimental, university operated station which went on the air in 1920 with a regular series of broadcasts of musical selections, which preceeded KDKA by several weeks. In 1920 the station broadcast the presidential elections.

WELLESLEY COLLEGE, WELLESLEY, MASS.
(Station WBS)

On April 20, 1942, in an extra edition of the Wellesley College News, it was announced that "Wellesley will hear the Voice of WBS Today".

The Wellesley News reported: "Following the opening bars of 'America the Beautiful', Rosamond Wilfley '42 will 'break the bottle over the mike' of Wellesley's new broadcasting station, WBS, 730 on the dial, today at 5:00 p.m. and she will introduce President Mildred H. McAfee, who will make a short speech on the value of a college radio system. The remainder of the program will be devoted to giving samples of the type of entertainment to be presented over the station. A radio play, 'Hunk is Punk' directed by Ruth Nagel '42, daughter of Conrad Nagel, will feature college talent.

"Following the skit, there will be a five minute summary of the program 'Boston This Weekend' which will be a feature of Friday broadcasts. Then Vasanthi Rama Rau '45 will give an analysis of the news. The time will be announced at 5:30 p.m. which will be followed by fifteen minutes of recorded music from George Gershwin's 'Porgy and Bess'. Murrayl Groh '42 will comment on the records.

"The grand finale of the first college broadcast, which can be heard all over the campus with the exception of Homestead and the Village houses, will feature a quiz show, the weekly variety program under the direction of Joan Davis '43 and the 'Battle of Wits' between the four classes. Taking part in the 'Battle of Wits' will be Suzanne Hayward '42, Patricia Wellington '43, Ann Crolius '44 and Lucile Peterson '45."

Station WBS was made possible by the enthusiastic initiative of several students and the generous financial backing of

Mr. Hill Blackett of Chicago, father of Priscilla Blackett '42. With the help of Miss McDowell, Professor of Physics, Wellesley is the first woman's college to set up independently a college station." So reported The Wellesley News.

CONGRATULATIONS

to

Wellesley College

on

the inauguration of

its

CAMPUS STATION

~~~~~~~~~~~~~~~~~~

The Yankee Network

and its

FM Stations

W38B	W43B
Mt. Washington	**Boston**

YALE UNIVERSITY, NEW HAVEN, CONN.
(Station WOCD)

Station WOCD, Yale University, commenced broadcasting on October 7, 1941. The station was owned and operated by The Yale Daily News and the call letters advertised the fact that The Yale Daily News is the nation's oldest daily college newspaper. By December of 1941, WOCD had forty men on its board including 14 announcers, 9 program monitors, 2 technicians, 5 business, and 10 script writers. Programs were broadcast from 5:00 p.m. to 9:45 p.m. and soon a morning schedule was also offered between 7:00 a.m. and 8:15 a.m. "designed to wake Yale students for morning classes." The program was heard daily except Sundays. Music, news and weather announcements were included in the morning broadcasts.

The Yale station WOCD was responsible for breaking up one of the largest race-bet rings in the State of Connecticut. The New York Herald Tribune reported as follows: "The Yale University undergraduate newspaper, 'The Yale Daily News', used an inch-and-a-half high streamer headline and a two column width story today to announce that its staff members and members of its radio station WOCD, have assisted police in breaking up a horse betting ring with headquarters on the edge of the university.

The headline, stretching across the five-column width of the paper read: *Elis Uncover Bookie Ring.* Underneath was a two-column sub-headline reading: *Tip From WOCD Leads To Raid By New Haven Police Yesterday.* And under this was another sub-headline reading: *Authorities Claim That Yale Men's Undercover Work May Reveal State Wide Gambling Racket.*

The story, in two-column measure, went on to relate how police, accompanied by undergraduates, swooped down on three rooms in Chapel Street, a block from The Daily News building and found betting records, direct wire telephones, head sets and other evidence of bookie operations. Seven persons were arrested in the raids on this and other places.

The story continued: "Entire credit for the information which led to the raid is due to Yale's WOCD broadcasting station's being able to expose the ring through a freak phone connection in its technical facilities. About a month ago student broadcasters began hearing reports of betting odds and names of horses trickling into the station. They called this to the attention of the police."

WOCD became a major extra-curricular activity at Yale. Today the station operates as The Yale Broadcasting Co., Inc. and is a commercial station with a power of 1,200 watts.

APPENDIX I

IBS MEMBER STATIONS PRIOR TO 1946

COLLEGE   MEMBERSHIP DATE

University of Alabama ....Sept. 1942
Antioch College .... March 31, 1942
Brown University
    Pembroke College...................... Feb. 18, 1940
Bryn Mawr College.......................... Oct. 1945
Bucknell University ....March, 1945
University of Connecticut .... Feb. 1940
Cornell University .... Feb. 18, 1940
Columbia University
    Barnard College .... April 11, 1941
UCLA .... Nov. 7, 1941
Harvard University...................... Dec. 28, 1940
Hamilton College........................June 22, 1942
Haverford College.......................... June, 1943
Knox College ....June 22, 1942
Mary Washington College................. Oct. 11, 1945
University of Maryland .... April 3, 1942
Ohio University .... July 30, 1943
Ohio State University.......................May, 1942
University of Pennsylvania .... Dec. 19, 1945
Princeton University .... Dec. 12, 1940
Radcliffe College............................ Oct. 1943
Rhode Island State College .... Feb. 18, 1940
Stephens College........................ April 26, 1944
Swarthmore College .... Dec. 28, 1940
Union College .... Dec. 28, 1940
Wellesley College .... Feb. 18, 1940
Williams College .... Feb. 18, 1940
Yale University .... Sept. 18, 1941

Some member stations commenced broadcasting prior to
the formation of IBS. Check Chapter IX, The Pioneer Stations.

IBS TRIAL MEMBER STATIONS PRIOR TO 1946

COLLEGE   MEMBERSHIP DATE

Brigham Young University	Feb. 22, 1943
Brooklyn College	May, 1942
University of Calif.-Berkeley	Nov. 5, 1941
Coe College	Nov. 1941
Colorado State College	April 15, 1941
University of Colorado	Nov. 1942
Dartmouth College	Dec. 28, 1940
Emerson College	Nov. 22, 1942
Emory University	Dec. 1943
University of Florida	Oct. 1941
Georgetown University	Sept. 1941
Hampton Institute	Dec. 28, 1940
Iowa State University	March 3, 1941
Mac Murray College	July 4, 1944
Massachusetts State	April, 1942
University of Michigan	Nov. 1942
University of Mississippi	Sept. 15, 1942
Montana State University	Feb. 4, 1946
University of Nebraska	Nov. 1942
University of North Carolina	April, 1942
North Carolina State	Dec. 21. 1945
Oklahoma A. & M.	Nov. 1941
College of The Pacific	March 20, 1946
St. Lawrence University	Oct. 1945
University of Southern Calif.	March, 1942
University of South Carolina	Dec. 1945
Stanford University	Sept. 1, 1941
Syracuse University	June 17, 1942

Trial Member stations are those which were in the planning stage or under construction.

APPENDIX II

RESULTS OF THE SURVEY TAKEN IN APRIL, 1945 TO DETERMINE THE LISTENERSHIP TO THE GENERAL ELECTRIC CAMPUS NEWS PROGRAM

Interviewing 591 students on seven campuses chosen at random, the results of the GE survey were as follows:

1. Question: Do you have access to a radio

	YES
Alabama	98%
Cornell	97%
Haverford	92%
Radcliffe	97%
Union	87%
Williams	96%
Yale	93%

Question 2: What company sponsors the campus news programs over your college station?

	Percentage correctly identifying General Electric Company as the sponsor
Alabama	78%
Cornell	77%
Haverford	87%
Radcliffe	49%
Union	98%
Williams	82%
Yale	60%

3. In spite of the fact that college stations broadcast only 2-6 hours daily while standard stations are on the air 16-24 hours per day, we asked the following question:

What stations do you listen to most often?

Here are the results listed in order of preference:

Yale
WOCD (College Station)
WELI
WNHC
WEAF
WABC
WOR
WTIC

Cornell
CRG (College Station)
WHCU
WHAM
WSYR
WAGE
WJZ
WABC

Williams
WMS (College Station)
WGY
WABC
WTRY
WEAF
WOR
WJZ
WBRK

Alabama
WJRD
WAPI
BRN (College Station)
WSGN
WBRC
WWL
WLWL

Radcliffe
WCOP
WEEI
Radio Radcliffe (College Station)
WBZ
WHDH
WNAC
WORL
WMEX

Union
WGY
WSNY
WTRY
UCRS (College Station)
WOKO
WABY
WABC

APPENDIX III

IBS PROMOTION BROCHURE "FROM PRINCETON TO STANFORD IBS SELLS THE COLLEGES"

This is the story of Toby Green, Williams '43

A boy with well-draped clothes and a future

Student of Ec and Statistics

so he can go into the family business

A boy who spends money

and is going to keep on spending it

A boy you'll want to know more about

TOBY GREEN and most undergraduates buy over One Billion dollars of merchandise, services, and entertainment in each college year. They come from families with means. Individually, Toby himself spends more from September to June than the Average American Family spends yearly.

TOBY GREEN and his friends set styles and habits, not just for youth, but for the whole country. Remember what happened when Toby changed to a pork-pie hat? Or think of the fever that grips the country when football gets under way in September. Newspapers and magazines watch Toby closely. They mirror his likes and dislikes. All over the country, men and women, young and old, follow suit.

TOBY is a boy who's going places. He'll be a leader in business, in society. Thousands of other undergraduates will become known as judges, accountants, engineers, professors and executives. They will be looked up to by others. Their tastes and habits will guide the people around them.

TOBY can be a mighty important boy to you.

He is still young -

His ideas are pliable -

If you can reach an audience of Toby Green and the
thousands of undergraduates like him,

If you can mould and create his attitudes and
tastes ....

YOU are not only reaching the college audience,

But a far greater audience of people who follow the
styles that colleges set ....

Of people who follow the leadership of Toby Green
and his friends, whether they go into
business or a profession.

If you want to know how to reach Toby Green, follow him on one of his typical days ...

at Williams

TOBY comes back to his room every afternoon about five -- worn out by a game of squash or an afternoon in the Lab. His first thought -- relaxation. He gets it with a turn of his radio dial. Toby and the majority of Williams undergraduates usually listen to their radios in the afternoon and after dinner, but most of all between 5 and 7 each evening. They rarely listen alone, for group listening is an important habit at college. Like Toby and the rest of Williams, each college has its own special listening times -- its "peak audiences" when the majority of undergraduates are tuned to their favorite station.

For example:*

COLLEGE	PEAK LISTEN- ING TIME
Cornell	*8 - 11 P.M.*
Williams	*5 - 6:15 P.M.*
Princeton	*5 - 6 P.M.*
Union	*7 - 8:15 P.M.*
Columbia	*9 - 11 P.M.*
Connecticut	*6 - 7:30 P.M.*
Wesleyan	*4:30 - 6 P.M.*

* All figures used are from the Intercollegiate Broadcasting System survey made at seven Eastern colleges between December 11 and 15.

PRACTICALLY every one of Toby's friends has a radio in his room.
The few who haven't never fail to join Toby for a good hour of
listening in his room or at the fraternity. 86.5% of Williams
students interviewed had their own sets. 100% had access to a
radio. Like proms, football and "bull sessions", radio has become a "must" in college life.

Percentage of all college students interviewed who have their own radios   84.3%

Percentage of all college students interviewed who have access to a radio   98.4%

TOBY GREEN like most undergraduates prefers the college station above all others.

It is something new -- an idea which has spread across the country from Princeton to Stanford in five years.

A station just for the colleges with studios on the campus.

Its staff made up of undergraduates like Toby.

Their own station...

Appealing to their tastes and interests —

The latest jazz,

The best in classical music,

Drama, news, forums,

Sports and comedy — the way they want it.

That is why the majority of undergraduates prefer the college station above all others.

The results of the College Radio Survey taken on December 11 and December 15, 1941, at two different hours show that of those students who were listening to their radios:

College	Percent Listening to College Station	Percent Listening to All Other Stations
Princeton	65.3%	34.7%
Wesleyan	91 %	9 %
Union	44.5%	55.5%
Connecticut	44.4%	55.6%
Williams	91.6%	8.4%
Cornell	60.4%	39.6%
Columbia	12.8%	87.2%

TOBY and his friends have formed the habit of leaving their
dials set to WMS . . .

Because they know they will get the kind of programs they want --

Programs which they can not only listen to, but study to.

Williams students listen an average of 1.52 hours daily to all
and every radio station. And of this time, they listen
as much as one hour daily to their own station WMS.
At all other colleges surveyed, undergraduates showed
a similar preference for their college stations.

Average time college stations are on the air each day:	4.74 hrs.
Average daily time spent listening to college stations by undergraduates interviewed:	1.26 hrs.
Average time network stations are on the air each day:	19-24 hrs.
Average daily time spent listening to network stations by undergraduates interviewed:	1.2 hrs.

LIKE Toby Green, thousands of undergraduates all over the United
States have come to depend on their college stations above all others.

Brown opened the first station in 1936.

When Wesleyan and Williams and Cornell followed, the Intercollegiate Broadcasting System began to grow like a
snowball.

During the Spring of 1941, over sixty colleges applied for membership . . . Stanford on the Coast, Antioch in the
Middle West, University of Florida in the South.

The IBS kept growing and growing till it dotted the country. On
December 30, 1941, its stations included:

Member Stations		Trial Members
Brown	*Pembroke*	*Alabama*
U. of California	*Princeton*	*Antioch*
U. C. L. A.	*Rhode Island State*	*Coe*
Columbia	*St. Edwards U.*	*Colorado State Col-* *lege of Education*
U. of Connecticut	*St. Lawrence**	*Dartmouth*
Cornell	*Swarthmore**	*Michigan*
U. of Florida	*Union*	*Oklahoma A. & M.*
Georgetown	*Wesleyan*	*Iowa State*
Hampton	*Williams*	*U. of Pennsylvania*
*Harvard**	*Yale*	*Stanford*
Knox		*Wellesley*

* Does not carry commercial programs

TO YOU -- the advertiser,

Who wants to reach Toby Green and

Thousands of American undergraduates

Who make up the richest and

The most untouched market in the United States . . . .

With stations at seventeen* colleges, the INTERCOLLEGIATE BROAD-
CASTING SYSTEM covers that market from Princeton to the University of California.

INDEX

INDEX

INDEX

North American Newspaper Alliance, 41
North Carolina State University, 110
North Carolina, University of, 110

Ohio State University, 109
Ohio University, 39, 94, 109
Oklahoma A. & M., 110, 120
Old Gold, 71
Oliphant Charles W., 89
O'Leary, Rev. Arthur A., 86

Page, Raymond, 68
Palmer, Frederick, 92
Parker, Malcom, 26
Parnicky, Joseph, J., 22, 26
Pembroke College, 26, 41, 109, 120
Pennsylvania, University of, 109, 120
Perkins, Warren, 103
Perry, John, 28
Peterson, Herb, 82
Peterson, Lucille, 104
Phoenix, Ronnie, 97
Pihl, Ted. 93
Pope Pius XII, 86
Princeton University, 40-42, 63, 70, 74, 95, 96, 109, 116, 118, 120
Providence Journal, 19
Providence, R. I., 27, 42, 64

R. J. Reynolds Tobacco Co., 42, 70, 82
Radcliffe College, 74, 97, 109, 111, 112
Radcliffe News, 97
Radio Daily, 36, 63
Radio Executives Club of New York, 43, 69
Radio Radcliffe (Radcliffe College), 97, 98
Rama Rau, Vasanthi, 104
RCA Victor, 71
Readers Digest, 42, 70
Reynolds, George, 68
Rhode Island State University, 26, 28, 39, 41, 63, 99, 109, 120
Richard, Prof. James, 81
Robinson, James, L., 95
Rockefeller, John D. Jr., 32
Roos, Bill, 81
Roosevelt, Eleanor, 49, 60-62

St. Lawrence University, 110, 120
Sanger, Mrs. Elliott, 56
San Paulo University (Brazil), 48
Sarrett, Tony, 82
Sarnoff, David, 18, 23
Sarnoff, Edward, 19, 23
Saturday Evening Post, 43, 44
Scheiner, Martin, 57

Scott-Smith, Edward, 82
Sondheim, James, 56
Siegelman, Abraham, 68
South Carolina, University of, 110
Southern California, University of, 110
Sprague, Sheldon, 68
Standard Oil Company of N. J., 36, 40, 65
Stanford Ink, 40
Stanford University, 42, 110, 120
Stasukevitch, John, 99
Stephens College, 73, 74, 109
Strauss, Jerome F. Jr., 14, 33
Stuart, Robert, 21
Studebaker, John W., 49, 74, 75
Stump, Prof. W. Turner, 27
Swarthmore College, 39, 100, 101, 109, 120
Syracuse University, 110

Taylor, Paul F., 22
Temkin, Edward, 82
Theis, H. Grant, 95, 96
Thomas, Helen M., 20
Thorpe, Peter, 17, 26
Tide Magazine, 36
Titus, Horace, 69
Travers, H. Linus, 25
Trolander, Hardy, 80
Tyng, William W., 89

UBS, UCRS, (Union College), 103, 112
UCBS (University of Connecticut), 82, 83
UCLA, 109, 120
Union College, 42, 63, 73, 74, 102, 103, 109, 111, 116, 118, 120

Van Dyke, Ruth, 21
Variety Magazine, 40

W2XMN (FM Station), Alpine, N. J., 54, 63
W2XQ (Radio Station), Union College, 103
W39B (FM Station), Mt. Washington, N. H., 63
W43B (FM Station), Boston, Mass., 63
W47NY (FM Station), New York, N.Y., 63
W57B (FM Station), Schnectady, N.Y., 63
W65H (FM Station), Hartford, Conn., 63
W71NY (FM Station), New York, N.Y., 54

INDEX